Student Study Guide

The Living World

GEORGE B. JOHNSON
Washington University, St. Louis

Prepared by

Lisa Shimeld
Crafton Hills College

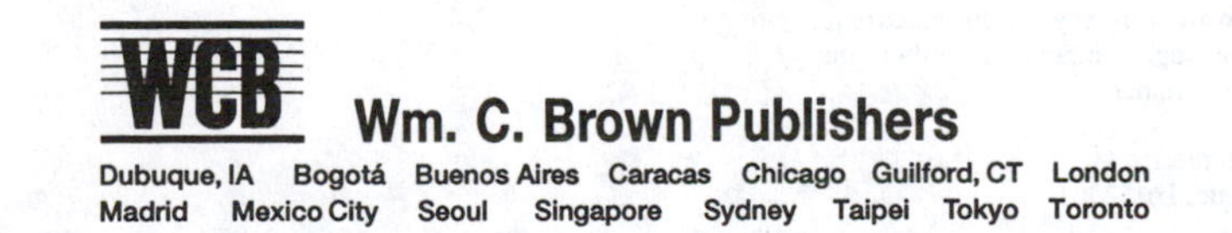

Wm. C. Brown Publishers
Dubuque, IA Bogotá Buenos Aires Caracas Chicago Guilford, CT London
Madrid Mexico City Seoul Singapore Sydney Taipei Tokyo Toronto

ISBN 0–697–22231–4

Printed in the United States of America
2460 Kerper Boulevard, Dubuque, IA 52001

10 9 8 7 6 5 4 3 2 1

Contents

Preface v

1 The Science of Biology 1

2 The Chemistry of Life 6

3 Cells 12

4 The Living Cell 17

5 Energy and Life 22

6 Foundations of Genetics 26

7 How Genes Work 31

8 Gene Technology 36

9 Evolution and Natural Selection 39

10 How We Name Living Things 44

11 The First Single-Celled Creatures 49

12 Advent of the Eukaryotes 53

13 Evolution of Multicellular Life 58

14 Rise of the Flowering Plants 63

15 The Living Plant 68

16 Evolution of the Animal Phyla 73

17 History of Terrestrial Vertebrates 78

18 How Humans Evolved 83

19 The Human Body 87

20 Circulation and Respiration 91

21 The Path of Food Through the Body 97

22 How the Body Defends Itself 102

23 The Nervous System 107

24 Chemical Signaling Within the Body 111

25 Human Reproduction and Development 115

26 Ecosystems 119

27 Living in Ecosystems 124

28 Planet Under Stress 129

Appendix: Answers to Chapter Questions 131

Preface

This study guide was written to accompany *The Living World* by George Johnson. Each of the 28 chapters contains a chapter overview, review questions (multiple choice, completion, and/or matching), and addresses for related web sites. The chapters also include art labeling exercises, key word searches, and crossword puzzles.

These learning aids will help students focus on the concepts introduced in each chapter of *The Living World.* The review questions enable students to double-check their knowledge of the material and prepare for exams. The art labeling exercises challenge students to recall what they have seen and read in the main text. The word searches and crossword puzzles are meant to be fun as well as to provide additional exposure to key terms and concepts. Answers to all questions and activities may be found in the Appendix.

1 The Science of Biology

Chapter Concepts

The discovery of how CFCs are reducing levels of ozone in the atmosphere is a good example of science in action. The scientific process is founded on careful observation. In a control experiment, only one variable is allowed to change. Scientific progress is made by rejecting hypotheses that are inconsistent with observation.

The acceptance of a hypothesis is always provisional. Well-tested hypotheses are often combined into general statements called theories. There is no surefire way to do science, no foolproof "method." One of the most creative aspects of scientific investigation is the formulation of novel hypotheses.

All living things share eight fundamental properties: complexity, movement, sensitivity, cellular organization, metabolism, homeostasis, reproduction, and heredity.

There are many ways to study biology. Six general themes often used to organize the study of biology are: levels of organization, the flow of energy, evolution, cooperation, structure determines function, and homeostasis.

Multiple Choice

1. The science of biology
   a. explores the magical processes by which our world operates.
   b. is conducted by people with poor organizational skills.
   c. discovers the general principles of our world by applying inductive reasoning.
   d. explores a variety of problems using the "gut" instincts of scientists.
   e. both a and d.

2. Using an algebraic formula to calculate the load capacity of a bridge is an example of
   a. deductive reasoning.
   b. inductive reasoning.
   c. theoretical math.
   d. applied math.
   e. mathematical reasoning.

3. Most new cars are manufactured with air conditioners that no longer use chlorofluorocarbons (CFCs) as coolants. This is in part due to
   a. their role in the reduction of light intensity in the atmosphere.
   b. their presence in the upper atmosphere for over 100 years.
   c. the allergic reaction that occurs in most people when they inhale CFCs.
   d. their role in the destruction of the ozone layer.
   e. both b and d.

4. Which of the following are in the correct order?
   a. Test, make observations, develop a hypothesis, make predictions, draw conclusions.
   b. Make observations, develop a hypothesis, make predictions, test, draw conclusions.
   c. Develop a hypothesis, test, make predictions, make observations, draw conclusions.
   d. Make predictions, develop a hypothesis, test, make observations, draw conclusions.
   e. Either a or b.

5. In an effort to improve crop yield a rancher tries different kinds of fertilizers. She uses an organic fertilizer on one field and a chemical fertilizer on another. A third field is not fertilized. The two fertilizers represent what part of her experiment?
   a. the hypothesis
   b. the control
   c. variables
   d. predictions
   e. a theory

6. The unfertilized field in the above experiment represents what part of her experiment?
   a. the hypothesis
   b. the control
   c. variables
   d. predictions
   e. a theory

7. The scientific method is limited to
   a. events that occur in our solar system.
   b. supernatural phenomena.
   c. religious questions.
   d. mathematical problems.
   e. observable and measurable processes.

8. A group of individuals living together is referred to as a(n)
   a. population.
   b. community.
   c. species.
   d. ecosystem.
   e. biome.

9. A tropical forest is an example of a(n)
   a. population.
   b. community.
   c. species.
   d. ecosystem.
   e. biome.

10. While walking through a deciduous forest in New England one might observe squirrels, blue jays, gypsy moths, sugar maples, and blue spruce. All of those organisms represent a(n)
    a. population.
    b. community.
    c. species.
    d. ecosystem.
    e. biome.

11. A bacterial colony consists of identical cells that are derived from a single original cell. Those cells represent members of the same
    a. population.
    b. community.
    c. species.
    d. ecosystem.
    e. biome.

12. Over 300 breeds of purebred dogs exist today. They are the result of
    a. evolution.
    b. natural selection.
    c. artificial selection.
    d. selective selection.
    e. adaptive evolution.

13. Scientists believe that the ozone layer over northern Europe may begin to break down because
    a. they have detected elevated levels of chlorine in the upper atmosphere in that area.
    b. the number of cases of skin cancer in that area has increased in recent years.
    c. crop yields are down in the area due to increased levels of ultraviolet radiation.
    d. that is the next likely place to be affected after Antarctica.
    e. none of the above.

14. Ultraviolet light causes skin cancer by
    a. damaging the organelles in skin cells resulting in cell death.
    b. damaging the DNA in skin cells.
    c. increasing the quantity of melanin in skin cells.
    d. concentrating toxins in skin cells.
    e. both a and c.

Completion

1. In the scientific method an educated guess as to the answer of a question is referred to as a(n)__________ .
2. __________ is the use of energy by living things.
3. The genetic systems seen in all living organisms are based on a molecule called __________ . This molecule is organized into units called __________ .
4. __________ are made of cells sharing similar structure and function.
5. An ecosystem is made up of __________ and __________ .
6. __________ wrote the book __________ in which he discussed his theory of evolution by natural selection.

Cool Places on the 'Net

The American Institute of Biological Sciences can be contacted for information concerning environmental issues and other topics related to biology at:

AIBS@GWUVM..GWU.EDU or at aibs@aibs.org

The Centers for Disease Control and Prevention home page can be found at:

http://www.cdc.gov

Key Word Search

M	P	E	P	S	I	S	E	H	T	O	P	Y	H	E
T	R	U	X	V	F	J	O	R	B	E	P	N	E	N
I	E	B	Y	P	E	Z	W	S	I	R	C	W	R	O
S	W	A	I	R	E	C	E	K	O	C	P	T	E	Z
S	C	C	T	J	O	R	H	T	M	O	R	R	D	O
U	M	T	U	N	V	E	I	Y	E	R	E	T	I	G
E	N	E	T	A	A	S	H	M	S	G	D	E	T	V
S	V	R	T	G	T	L	Z	T	E	A	I	I	Y	A
N	O	I	T	A	L	U	P	O	P	N	C	S	Z	R
L	O	A	T	E	B	W	M	I	R	E	T	N	T	I
N	W	P	C	C	Z	O	G	Z	P	L	I	A	I	A
S	L	V	W	M	U	N	L	T	E	L	O	G	Y	B
E	V	I	T	C	U	D	N	I	B	E	N	R	R	L
C	Y	P	M	F	V	M	E	T	S	Y	S	O	C	E
R	A	P	X	K	I	N	G	D	O	M	H	Q	B	O

Biomes
Cell
Control
Deductive
Ecosystem
Eubacteria
Experiment
Fungi

Heredity
Hypothesis
Inductive
Kingdom
Metabolism
Observation
Organelle
Organs

Ozone
Plantae
Population
Prediction
Protista
Theory
Tissues
Variable

Crossword

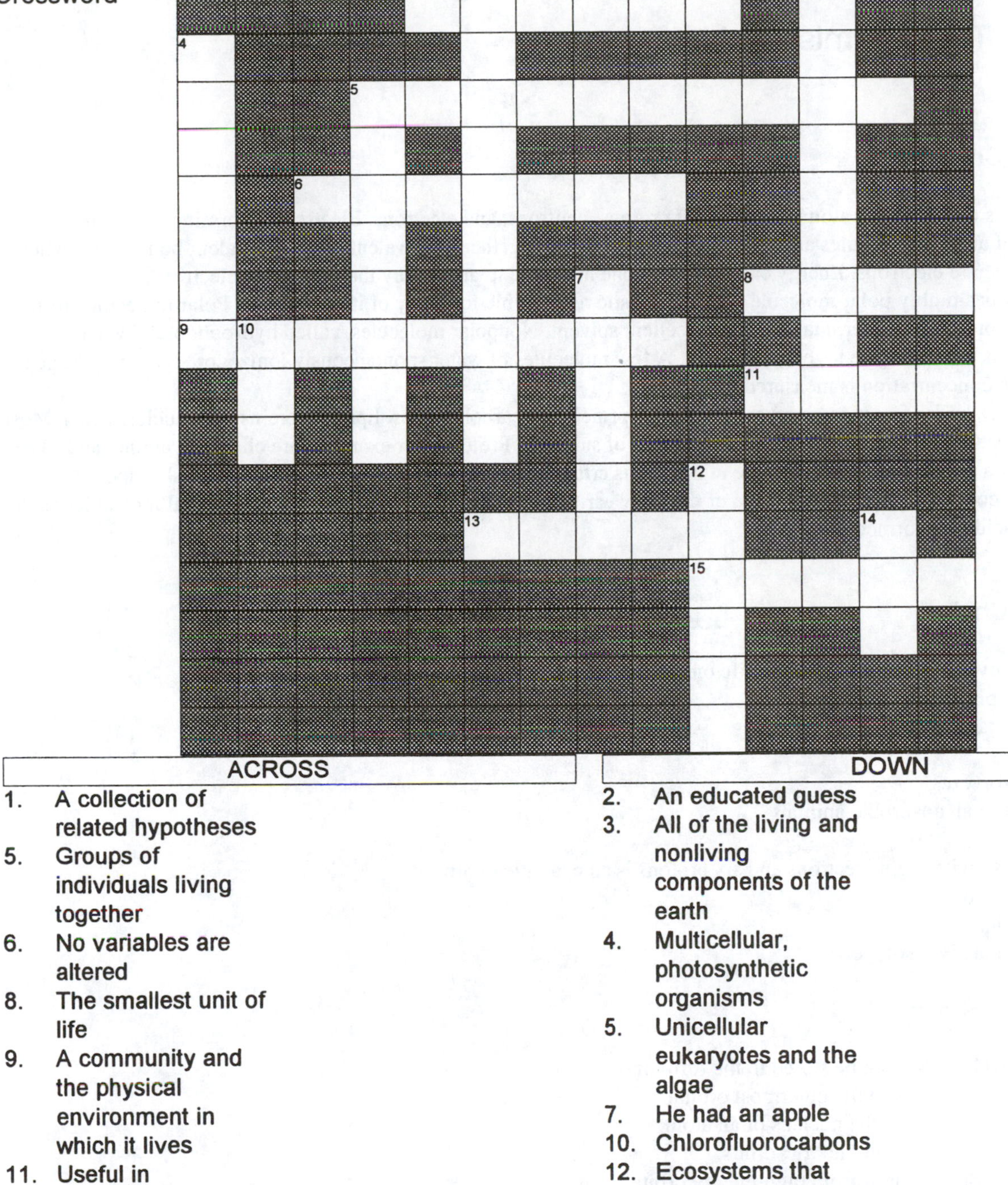

ACROSS

1. A collection of related hypotheses
5. Groups of individuals living together
6. No variables are altered
8. The smallest unit of life
9. A community and the physical environment in which it lives
11. Useful in experiments but hated by students
13. Kingdom that contains yeasts and molds
15. There is a big hole in this layer

DOWN

2. An educated guess
3. All of the living and nonliving components of the earth
4. Multicellular, photosynthetic organisms
5. Unicellular eukaryotes and the algae
7. He had an apple
10. Chlorofluorocarbons
12. Ecosystems that occur in several places worldwide, tropical rain forest is an example
14. Genetic molecule

2 The Chemistry of Life

Chapter Concepts

All matter is composed of atoms, made up of protons, neutrons, and electrons. Electrons determine the chemical behavior of atoms. Molecules are collections of atoms held together by covalent bonds. Covalent bonds form when two atoms share electrons. Energy can pass from one molecule to another by the transfer of electrons.

Water is a highly polar molecule, a characteristic responsible for many of its properties. Polar molecules form hydrogen bonds with water, making it an excellent solvent. Nonpolar molecules, called hydrophobic ("water-hating"), tend to aggregate together in water. A few molecules of water spontaneously ionize, producing hydrogen ions whose concentration is measured by pH.

Cells are built of very large macromolecules, principally carbohydrates, lipids, proteins, and nucleic acids. Most macromolecules are assembled by forming chains of subunits. Proteins, for example, are chains of amino acids.The function of a particular macromolecule in the cell is critically dependent upon its three-dimensional shape.

When conditions resembling those of the early earth are re-created in the laboratory, biological molecules such as amino acids are formed.

Multiple Choice

1. Fossil evidence suggests that all life on earth probably arose from which of the following?
   a. simple arthropods
   b. prokaryotic cells
   c. apes
   d. herbivores
   e. colonial, anaerobic animals

2. An atom with eight electrons and six protons is an example of a(n)
   a. molecule.
   b. mole.
   c. radioactive isotope.
   d. ion.
   e. polar element.

3. Chemical energy can be stored in the form of
   a. extra electrons in the outermost orbital.
   b. extra protons in the nucleus of an atom.
   c. electrons in "high-energy" orbits.
   d. electrons released from high-energy orbital.
   e. both a and c.

4. When a substance is oxidized in a cell another substance is __________ at the same time.
   a. removed
   b. oxidized
   c. hydrolyzed
   d. reduced
   e. lysed

5. __________ are the type of bond found in crystals such as table salt.
   a. Ionic bonds
   b. Covalent bonds
   c. Hydrogen bonds
   d. van der Waals
   e. Both a and d.

6. It takes quite a lot of energy to bring water to a boil. This is due to
   a. the covalent bonds in a molecule of water.
   b. the hydrogen bonds between adjacent water molecules.
   c. the ionic bonds that are broken when water reaches the boiling point.
   d. excessive quantities of hydrogen ion found in impure water.
   e. both a and d.

7. If an atom has four electrons in its second energy level then which of the following must be true?
   a. The third energy level must have at least two electrons.
   b. The third energy level must be empty.
   c. The first energy level must have two electrons.
   d. The first energy level must have four electrons.
   e. The first energy level must be empty.

8. A research chemist discovers a new element. She determines that the atoms of this element are particularly reactive. That means that the atom probably
   a. has an incompletely filled outer electron orbital.
   b. has a filled outer electron orbital.
   c. is radioactive.
   d. has a high specific heat.
   e. has probably been discovered before.

9. Ionic bonding involves
   a. the sharing of electrons between two or more atoms.
   b. weak bonding between adjacent salt molecules.
   c. weak bonding between adjacent water molecules.
   d. the attraction of oppositely charged atoms.
   e. both a and d.

10. One reason that ionic bonds do not play an important role in most biological molecules is that they are
    a. incompatible with water.
    b. directional.
    c. not directional.
    d. too strong to be broken in the cell.
    e. unable to form in carbon-based molecules.

11. Hydrogen bonding in water is a consequence of
    a. the polar covalent nature of water molecules.
    b. the nonpolar bonding that occurs between oxygen and hydrogen.
    c. excessive hydrogen dissolved in the water.
    d. both a and b.
    e. both a and c.

12. Hydrogen bonds are
    a. weak bonds that last only a fraction of a second.
    b. weak bonds that last indefinitely.
    c. weak bonds that hold molecules together.
    d. strong bonds that form between adjacent molecules.
    e. strong bonds that last for only a fraction of a second.

13. Sweating helps runners in a marathon sweat to cool off. This is because
    a. the water molecules in sweat adhere to the skin and have a cooling effect.
    b. for every gram of water that evaporates from their bodies, 586 calories of heat are removed.
    c. the water lost in sweating reduces body weight, making it easier for them to run.
    d. the salt in sweat has a cooling effect on the skin.
    e. both a and d.

14. A common medium used in the microbiology laboratory is carbohydrate fermentation broth. When introduced to the broth, a bacterium that is capable of fermenting the carbohydrate produces acidic waste products. What could the microbiologist add to the media to slow down changes in its pH?
    a. additional acid
    b. additional carbohydrate
    c. an additional source of nutrients
    d. a buffer
    e. a substance that is a strong base

15. Which of the following is a complex carbohydrate that is a component of fungi cell walls and insect exoskeletons?
    a. terpenes
    b. starch
    c. chitin
    d. steroids
    e. ATP

16. __________ are lipids that are a component of animal cell plasma membranes.
    a. Terpenes
    b. Starch
    c. Chitin
    d. Steroids
    e. ATP

17. __________ are long carbon chains that function as pigments in photosynthesis.
    a. Terpenes
    b. Starch
    c. Chitin
    d. Steroids
    e. ATP

18. Which of the following is a structural protein found in both birds and rhinos?
    a. glycogen
    b. cellulose
    c. chitin
    d. keratin
    e. none of the above

19. How do DNA and RNA differ?
    a. They have different ribose sugars.
    b. DNA is a double-stranded molecule while RNA is single-stranded.
    c. Thymine is present in DNA but not in RNA.
    d. Uracil is present in RNA but not in DNA.
    e. All of the above.

20. The bubble model suggests that
    a. the building blocks of life were generated inside of bubbles on the ocean's surface.
    b. bubbles of amino acids interacted to form more complex proteins on primordial earth.
    c. the first simple cells were actually bubbles of ocean water covered by a membrane.
    d. bubbles on the ocean's surface contained sufficient oxygen to facilitate chemical reactions.
    e. both a and d.

Completion

1. The three possible origins of life on earth are __________ , __________ , and __________ .
2. The attraction of like molecules is __________ , while the attraction of different substances is __________ .
3. An increase in the number of hydrogen ions in water causes it to become more __________ , while an increase in the number of hydroxide ions causes it to become more __________ .
4. Long chains of sugars are called __________ .
5. In DNA adenine pairs with __________ and cytosine pairs with __________ .

Cool Places on the 'Net

The Caltech Genome Research Laboratory can be contacted at:

http://www.tree.caltech.edu/

Label the Art

Identify each of the six functional groups according to their structural formulas.

Group	Structural Formula	Ball-and-Stick Model	Found In:
1. ______	—OH		Alcohols
2. ______	—C— ‖ O		Formaldehyde
3. ______	—C(=O)OH		Vinegar
4. ______	—N(H)H		Ammonia
5. ______	—S—H		Rubber
6. ______	—O—P(O⁻)(=O)—O⁻		ATP

Crossword

ACROSS

7. The attraction of different types of molecules
9. Deoxyribonucleic acid
10. Water fearing
11. Different forms of the same atom
14. A substance with more hydrogen ions than in pure water
15. Has an atomic number of 8
16. Type of bond where electrons are shared
19. _______ bonds join together amino acids to form proteins
21. Has an atomic number of 1
22. Positively or negatively charged atom
23. Negatively charged particle in atoms
25. A group of atoms bonded together

DOWN

1. Atomic _________ equals the protons + neutrons
2. A type of lipid found in biological membranes
3. Has an atomic number of 6
4. The gain of an electron
5. The attraction of like molecules
6. The loss of an electron
8. A substance with more hydroxyl ions than in water
12. Contains carbon and hydrogen, special groups are attached
13. Negatively charged particle found in oribitals surrounding the nucleus
17. DNA is this type of molecule
18. Water vapor
20. Type of bond formed by atoms with opposite charges
22. Solid form of water
24. Ribonucleic acid

3 Cells

Chapter Concepts

All life is composed of cells—about 100 trillion in the case of a human body. Cells tend to be quite small, limited by the distance to which substances must diffuse through the cytoplasm and also by the surface-to-volume ratio as cell size increases. Small cells have relatively more surface with which to interact with the environment.

The cell cytoplasm is bounded by the plasma membrane. The plasma membrane is a double layer of modified fat molecules called a lipid bilayer. Protein-lined passageways through the bilayer allow chemicals to enter and leave the cell.

The simplest cells are bacteria, called prokaryotes because they lack a nucleus. The key characteristic shared by all bacterial cells is that they lack internal compartments.

All nonbacterial cells contain a nucleus at some point and are called eukaryotes. The interior of eukaryotic cells is occupied by an extensive membrane system that creates many subcompartments within the cytoplasm. The cell's hereditary information, contained within chromosomes, is isolated within the nucleus. Proteins are assembled in the cytoplasm on complexes called ribosomes. Cells carry out oxidative metabolism within bacteria-like organelles called mitochondria.

Multiple Choice

1. Which of the following would be a problem for a cell if it were able to grow to twice its normal size?
   a. water retention
   b. salt balance
   c. reproduction
   d. intracellular communication
   e. communication between different types of cells

2. The cell theory states that
   a. cells arise spontaneously.
   b. cells arise from preexisting cells.
   c. cells are the smallest unit of life.
   d. organelles can survive when separated from cells.
   e. both b and c.

3. Which portion of a phospholipid molecule would be facing the environment outside of the cell?
   a. the polar head
   b. the nonpolar head
   c. the polar tail
   d. the nonpolar tail
   e. both a and d

4. Which portion of a phospholipid molecule would be facing the interior of the cell?
   a. the polar head
   b. the nonpolar head
   c. the polar tail
   d. the nonpolar tail
   e. both b and d

5. Which portion of a phospholipid molecule is repelled by water?
   a. the polar head
   b. the nonpolar head
   c. the polar tail
   d. the nonpolar tail
   e. none of the above

6. Membrane proteins
   a. exist in fixed positions within the plasma membrane.
   b. move freely within the plasma membrane.
   c. are involved in cell-to-cell recognition.
   d. are always large enough to span the entire width of the plasma membrane.
   e. are absent in animal cells.

7. How might viruses play a role in curing cystic fibrosis?
   a. Viruses could remove the defective gene from lung cells in persons with cystic fibrosis.
   b. Viruses could repair the defective gene in lung cells in persons with cystic fibrosis.
   c. Viruses could identify the defective gene for later removal.
   d. Viruses could deliver a copy of the normal gene to lung cells in persons with cystic fibrosis.
   e. Viruses cannot possibly play a role in curing cystic fibrosis.

8. Which of the following exhibit the 9 + 2 structure of microtubules?
   a. flagella
   b. cilia
   c. sensory hairs in the human ear
   d. basal body
   e. all of the above

9. Centrioles may have originated as
   a. a unique arrangement of microtubules found in some prokaryotic species.
   b. protein fibers produced by ribosomes in some eukaryotic cells.
   c. protozoans that formed an endosymbiotic relationship with primitive aerobic prokaryotes.
   d. spirochetes that became involved in an endosymbiotic relationship.
   e. locomotor structures.

10. Access to the nucleus is regulated by
    a. specialized proteins within the nuclear pore.
    b. the rate of osmosis.
    c. the amount of ATP available for active transport.
    d. hormones.
    e. the nucleolus.

Matching

1. _____ Ribosomes	a. Have their own DNA and are the site of oxidative metabolism
2. _____ Cell wall	b. The site of protein synthesis; made of RNA and protein
3. _____ Organelle	c. This structure surrounds the plasma membrane in plant cells
4. _____ Lysosome	d. A network of internal membranes
5. _____ Golgi complex	e. Form the cytoskeleton
6. _____ Endoplasmic reticulum	f. Function in the digestion of cell debris and cell death
7. _____ Mitochondrion	g. Plant organelle, contains photosynthetic pigments
8. _____ Chloroplast	h. Stacks of flattened vesicles
9. _____ Chromatin	i. Membrane-bounded structure with a specialized function
10. _____ Microtubule	j. Long, threadlike molecule of DNA

Completion

1. Human cells average between ________ and _______ micrometers in diameter.
2. As cells increase in size, the __________ increases at a faster rate than does the ___________ .
3. Persons having the disease called ____________ experience excessive mucus secretion that clogs the lungs.
4. Although animal cells lack a cell wall, their structure is maintained by the ___________ .
5. Microfilaments are made of the protein __________ , while microtubules are made of the protein ___________ .

Cool Places on the 'Net

To view electron micrographs of cells, check out on-line Laboratory Manual of Cell Biology and Histology from Emory University at:

http://arnica.csustan.edu/CE.html

For further information regarding cell biology, contact the American Society for Cell Biology at:

http://www.faseb.org/ascb/

Label the Art

A. Label the parts of this animal cell.

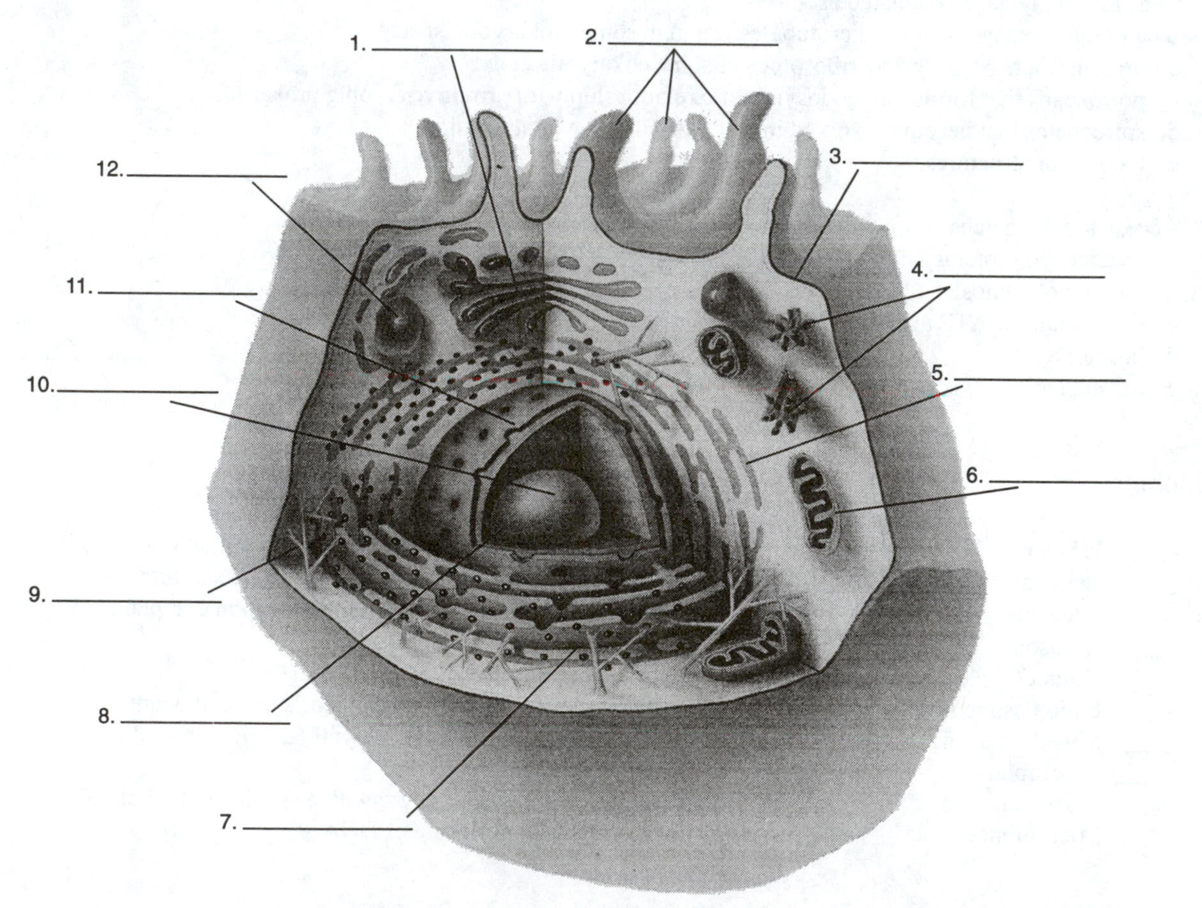

B. Label the parts of this chloroplast.

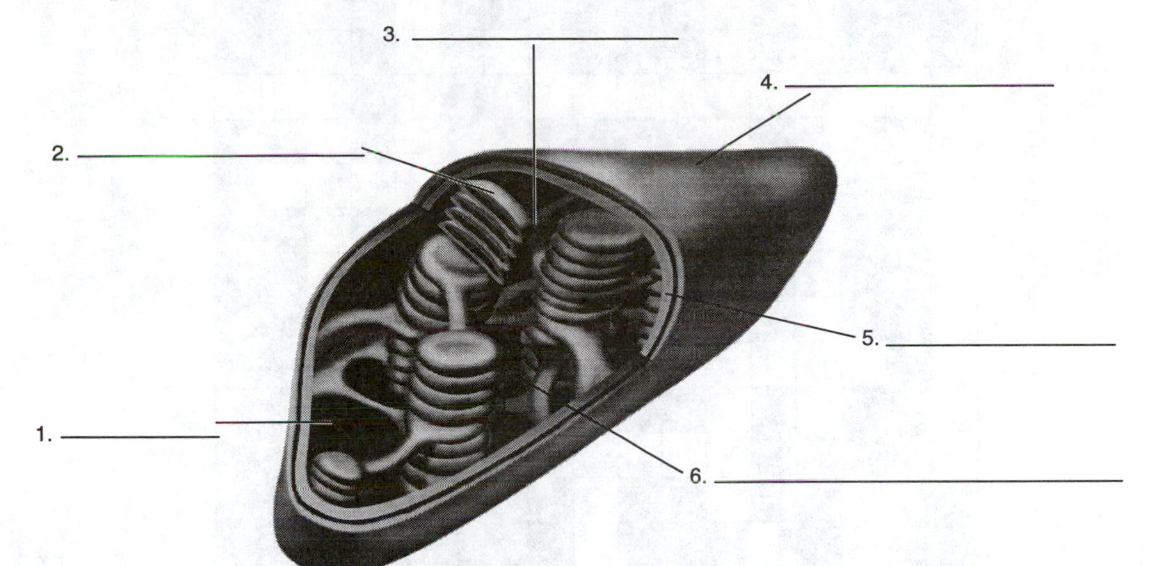

Key Word Search

E N A R B M E M A M S A L P D
L R O M S A L P O T Y C O H I
U I O T R C I L I A A A E O O
B B J P E H C J Y N Z H U S K
U O C L R L N H M D L C K P A
T S L H V O E U R V Z J A H L
O O S F R R K K C O T I R O Y
R M S T R O M A S L M C Y L H
C E E Y N P M I R O E A O I T
I B A S A L B O D Y T O T P G
M J T A N A R G S T O Y I I E
M L Y S O S O M E O W T C D N
C Y B H G T I U O Q M I I M E
Z W P E R O X I S O M E V C J
M U L L E G A L F C X J J Z A

Basal body
Cell
Chloroplast
Chromatin
Chromosome
Cilia
Cytoplasm
Cytoskeleton

DNA
Eukaryotic
Flagellum
Gene
Grana
Lysosome
Microtubule
Nucleoid

Peroxisome
Phospholipid
Plasma membrane
Prokaryotic
RNA
Ribosome
Stroma
Thylakoid

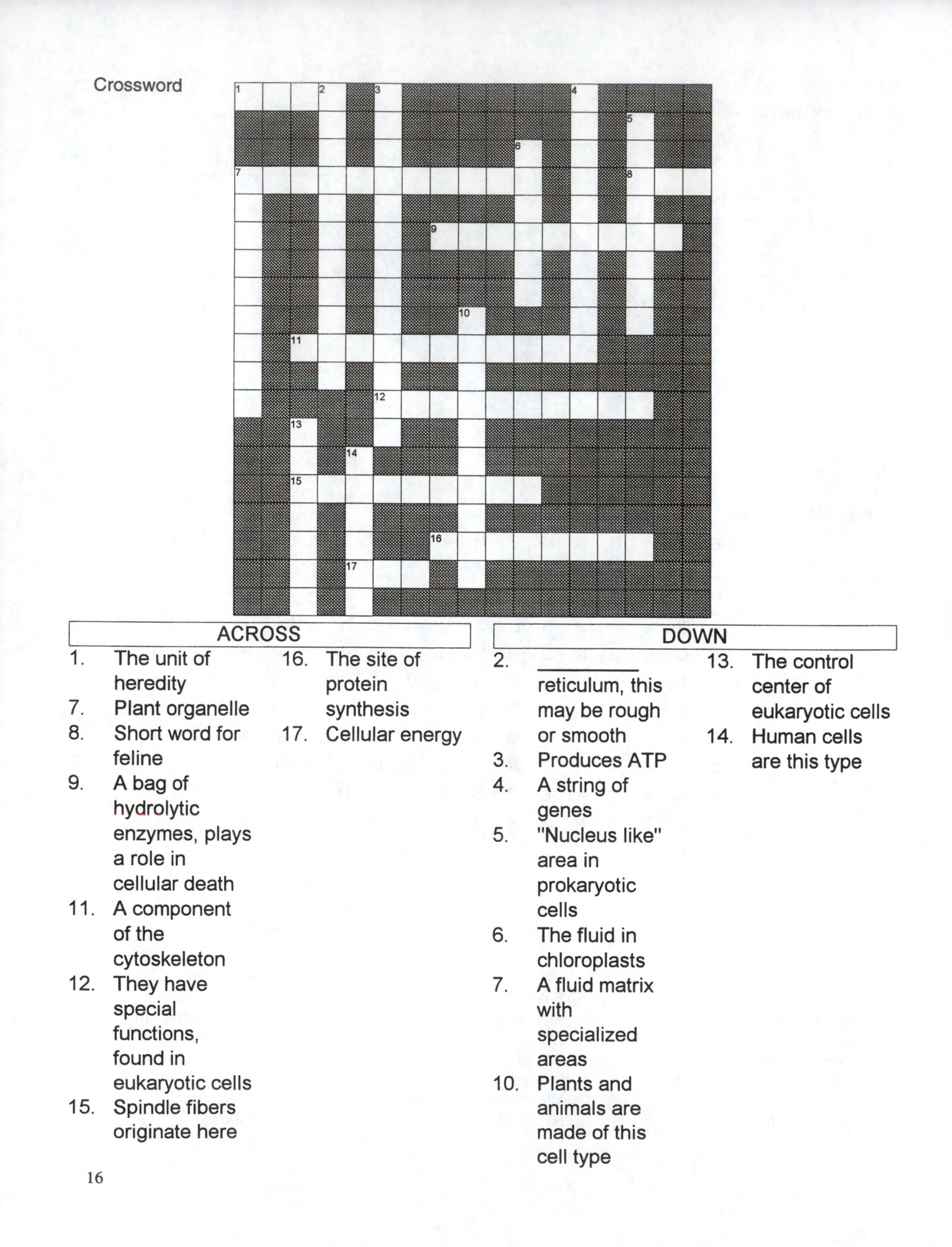

Crossword

ACROSS

1. The unit of heredity
7. Plant organelle
8. Short word for feline
9. A bag of hydrolytic enzymes, plays a role in cellular death
11. A component of the cytoskeleton
12. They have special functions, found in eukaryotic cells
15. Spindle fibers originate here
16. The site of protein synthesis
17. Cellular energy

DOWN

2. __________ reticulum, this may be rough or smooth
3. Produces ATP
4. A string of genes
5. "Nucleus like" area in prokaryotic cells
6. The fluid in chloroplasts
7. A fluid matrix with specialized areas
10. Plants and animals are made of this cell type
13. The control center of eukaryotic cells
14. Human cells are this type

4 The Living Cell

Chapter Concepts

All cells transport water and other molecules across their plasma membranes. Molecules move randomly through the cell by diffusion. Molecules diffuse down concentration gradients, thus tending to equalize concentrations. Osmosis occurs when water molecules are able to move through a membrane and other polar molecules are blocked. Selective transport of molecules can occur via facilitated diffusion or via active transport with the expenditure of ATP.

Receptor proteins protruding from the surface of cells receive chemical and electrical stimuli and transfer this information to the cell interior. Chemical signals called hormones pass information from one cell to another. Voltage-sensitive channels allow a cell to respond to changes in its electrical surroundings.

Bacterial cells divide by simply splitting into two halves. Eukaryotic cells divide by mitosis, a complex process that delivers one replica of each chromosome to each of the two daughter cells.

In humans, cells are programmed to undergo a limited number of cell divisions before dying.

Multiple Choice

1. A busy student pours herself a cup of coffee and adds a sugar cube. A moment later the phone rings so she doesn't have time to stir it. Ten minutes later she returns to her coffee, and when she tastes it, it is obvious that the sugar has dissolved and that the coffee is uniformly sweet. What happened?
   a. Osmosis has caused the sugar molecules to move down the concentration gradient until they were evenly distributed in the coffee.
   b. Facilitated diffusion has occurred and has caused the sugar molecules to dissolve in the coffee.
   c. Diffusion has occurred in the coffee causing the sugar to become evenly distributed.
   d. Active transport has occurred using the heat in the coffee as a source of energy.
   e. None of the above.

2. The sugar molecules in the coffee mentioned above bounce and jostle each other because they have
   a. kinetic energy.
   b. potential energy.
   c. static energy.
   d. random energy.
   e. occasional energy.

3. Water molecules crossing a plasma membrane do so by
   a. osmosis.
   b. diffusion.
   c. facilitated diffusion.
   d. active transport.
   e. transport by the sodium-potassium pump.

4. Which of the following is a requirement for osmosis to occur?
   a. a semipermeable membrane
   b. a concentration gradient
   c. sugar or salt molecules dissolved in water
   d. an energy source
   e. all of the above

5. What advantage would a plant cell have over an animal cell if both were placed in pure water?
   a. The plant cell would resist shrinking in pure water.
   b. The plant cell would resist bursting in pure water.
   c. The plant cell would reproduce at a faster rate in pure water.
   d. Both a and c.
   e. None of the above.

6. The release of insulin from certain human cells is an example of
   a. endocytosis.
   b. exocytosis.
   c. phagocytosis.
   d. pinocytosis.
   e. both a and c.

7. An amoeba is engulfing bacterial cells by surrounding them with a pseudopod. This is an example of
   a. endocytosis.
   b. exocytosis.
   c. phagocytosis.
   d. pinocytosis.
   e. both a and c.

8. The removal of waste products from the amoeba is an example of
   a. endocytosis.
   b. exocytosis.
   c. phagocytosis.
   d. pinocytosis.
   e. both a and c.

9. Diffusion of molecules across biological membranes that requires channels is called
   a. active transport.
   b. the sodium-potassium pump.
   c. facilitated diffusion.
   d. osmosis.
   e. none of the above.

10. The sodium-potassium pump functions to
    a. remove sodium and potassium ions from the cell.
    b. pump sodium and potassium ions into the cell.
    c. pump potassium ions into the cell and sodium ions out of the cell.
    d. pump sodium ions into the cell and potassium ions out of the cell.
    e. pump sodium and potassium ions into the cell and calcium out of the cell.

11. A liver cell would detect the presence of insulin by
    a. means of receptor proteins on its surface.
    b. means of receptor carbohydrates on its surface.
    c. producing insulin detection binding molecules.
    d. the reduction of glucose in the external environment.
    e. both a and d.

12. The structure of eukaryotic chromosomes is in part maintained by
    a. covalent bonding between the strands of the double helix.
    b. ionic bonding between the strands of the double helix.
    c. winding the DNA molecule around a core of histone protein.
    d. the centromere.
    e. microtubules present in the cytoplasm.

13. Sister chromatids are
    a. pairs of heterozygous chromosomes.
    b. only seen in prokaryotic cells.
    c. formed only during meiosis.
    d. duplicated DNA held together by a centromere.
    e. both a and d.

14. DNA binds with histone proteins because
    a. the combination of the two creates a positively charged structure.
    b. the positively charged histone is attracted to the negatively charged DNA.
    c. the combination of the two creates a complex with no net charge.
    d. that is the only way that DNA replication can be stimulated.
    e. none of the above.

15. Human skin cells are
    a. dead when mature.
    b. haploid.
    c. diploid.
    d. capable of becoming germ cells.
    e. produced by meiosis.

16. While viewing stained onion root tip cells you notice that distinct chromosomes cannot be observed in most of the cells. This is because most of the cells
    a. were destroyed while preparing the slide.
    b. must be in interphase.
    c. are in metaphase.
    d. must be undergoing meiosis.
    e. are in stasis.

17. The S phase of mitosis involves synthesis of
    a. cellular products for export from the cell.
    b. cellular membranes and organelles.
    c. DNA.
    d. the cell wall in plant cells.
    e. Both a and d.

18. The majority of the cell cycle is taken up by
    a. interphase.
    b. prophase.
    c. metaphase.
    d. anaphase.
    e. telophase.

19. The nuclear envelope reforms in the daughter cells during
    a. interphase.
    b. prophase.
    c. metaphase.
    d. anaphase.
    e. telophase.

20. The shortest stage of mitosis is
    a. interphase.
    b. prophase.
    c. metaphase.
    d. anaphase.
    e. telophase.

Completion

1. _________ is the movement of molecules across a membrane and against a concentration gradient. __________ is the energy used to fuel this process.
2. More than one-third of the energy expended by a human cell is used to drive the __________ .
3. Most bacterial cells reproduce by a method known as _______________ .

Cool Places on the 'Net

For information on current research in cell biology, try "Cell and Molecular Biology Online" at:

http://www.tiac.net/users/pmgannon/

Also of interest is the American Society for Cell Biology at:

http://www.faseb.org/ascb/

Label the Art

Label the stages of the cell cycle. (Note: These drawings have been deliberately arranged out of order.)

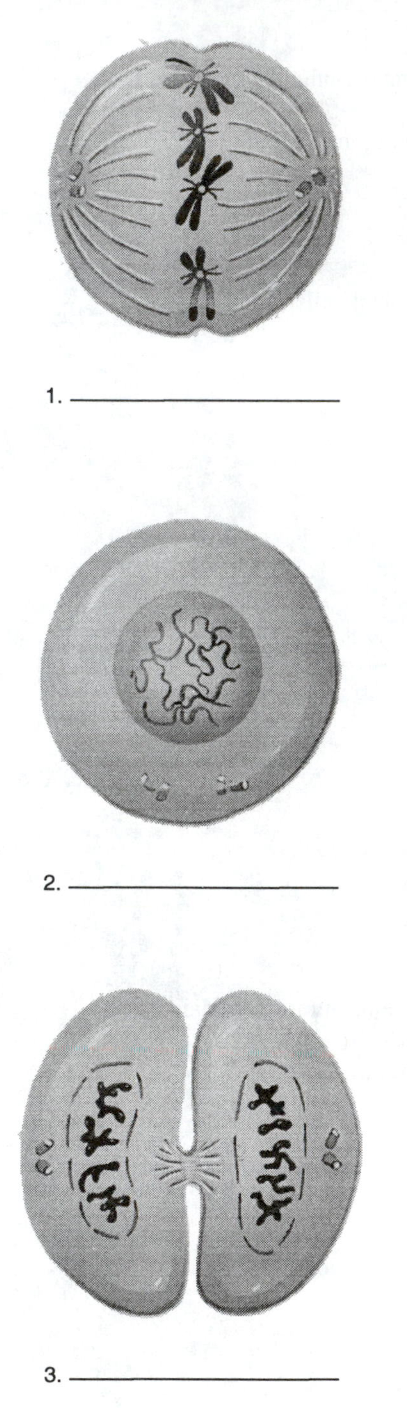

1. ______________________

2. ______________________

3. ______________________

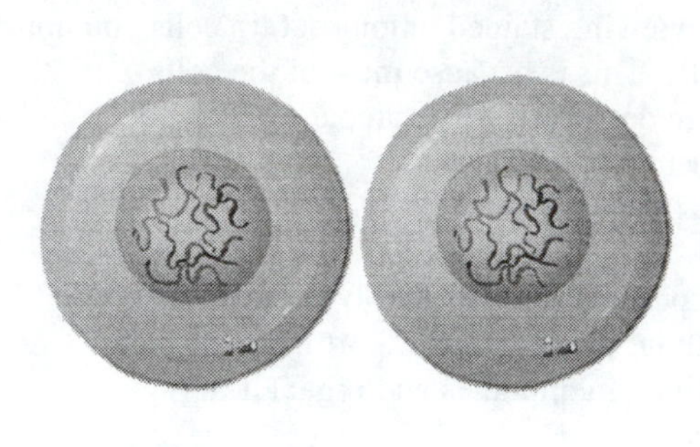

4. ______________________

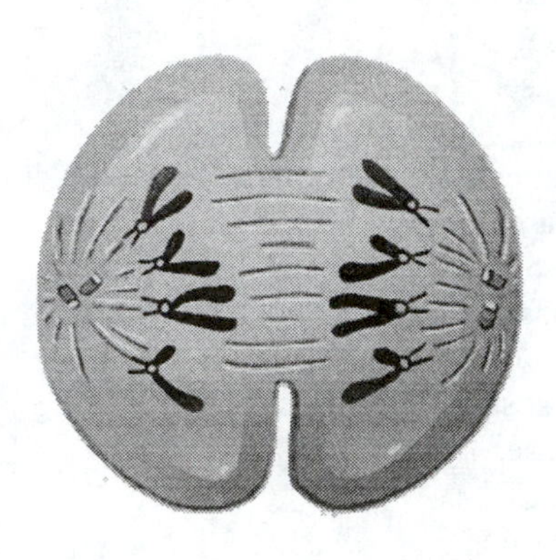

5. ______________________

6. ______________________

Crossword

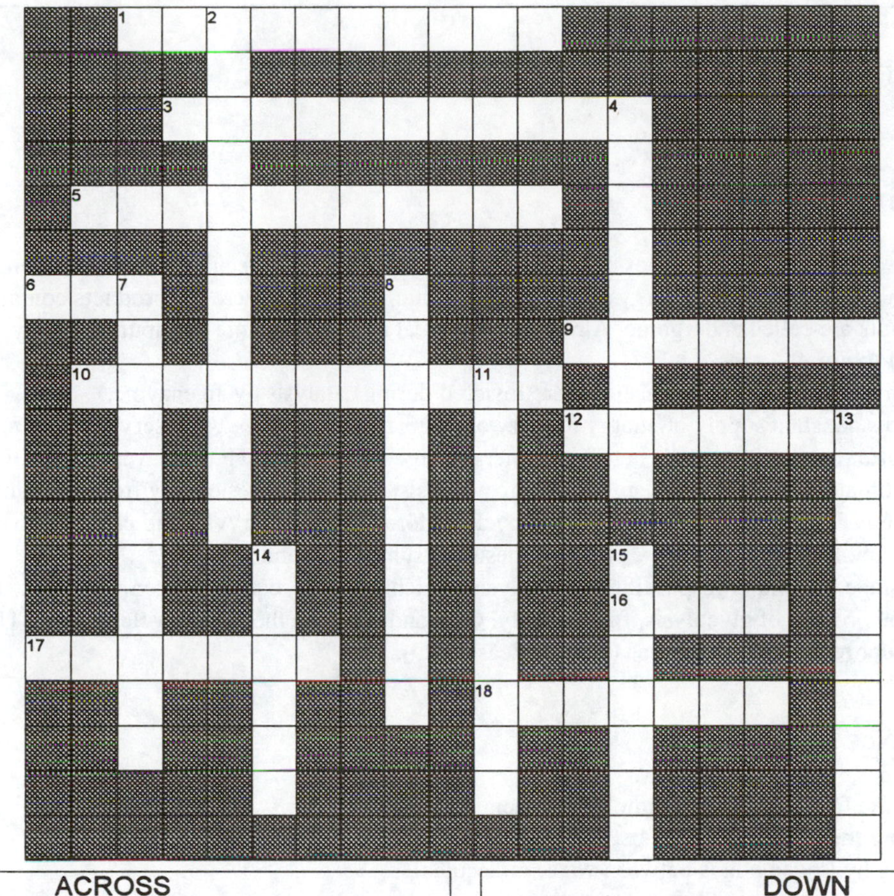

ACROSS	DOWN

ACROSS

1. 10% salt solution as opposed to human blood
3. ________ diffusion occurs through protein channels in the plasma membrane
5. Division of cytoplasm
6. "Energetic" molecule
9. Diffusion across a biological membrane
10. The movement of materials into the cell
12. Nonreproductive cells in humans
16. One end of a cell
17. Protein associated with eukaryotic DNA
18. A nucleotide in DNA

DOWN

2. Engulfing a large fragment of organic material by amoeba is by this process
4. Random movement of molecules down a concentration gradient
7. Cell "drinking"
8. The removal of substances from the cell
11. The period between cell divisions
13. In the middle of some chromosomes, a constriction
14. Bacteria reproduce by _______ fission
15. _______ fibers attach to chromosomes during mitosis

5 Energy and Life

Chapter Concepts

The chemistry of cells involves a series of chemical reactions between molecules in which a chemical bond is made or broken. Reactions that release energy are called exergonic, and those where the products contain more energy than the reactants are called endergonic. Almost all chemical reactions require an input of energy, called activation energy, to start them off.

Activation energies of cellular reactions are lowered during catalysis by an enzyme.

Cells need a constant supply of energy to carry out their many activities. ATP serves as the molecular energy currency for virtually all of the cell's activities. Energy is released when ATP is cleaved into ADP + P_i.

Energy ultimately reaches organisms via photosynthesis, which captures energy from sunlight and uses this energy to make ATP and NADPH. In the Calvin cycle, chloroplasts use the ATP and NADPH to convert CO_2 in the air into organic molecules. Ultimately, photosynthesis consumes CO_2 and releases O_2.

The first stage of cellular respiration is glycolysis, which does not require oxygen. The next stages involve oxidation of the product of glycolysis, first to acetyl CoA and then, via the Krebs cycle, to CO_2. Ultimately, oxidative cellular respiration consumes O_2 and releases CO_2.

Multiple Choice

1. Which of the following is an example of a chemical reaction?
   a. breaking the hydrogen bonds between water molecules
   b. forming hydrogen bonds between water molecules
   c. joining amino acids to form a protein molecule
   d. storing fat in cells
   e. all of the above

2. The energy required to initiate a chemical reaction is called
   a. activation energy.
   b. initiation energy.
   c. catalysis.
   d. primary energy.
   e. initial energy.

3. Even when catalysts are available, which of the following is also required for a reaction to occur?
   a. favorable atmospheric conditions
   b. oxygen
   c. energy
   d. zinc
   e. hydrogen

4. In many reactions NAD^+ functions to
   a. accept electrons.
   b. transform electrons.
   c. release electrons.
   d. all of the above.
   e. Both a and c.

5. A cell might regulate the production of a certain substance by
   a. destroying the enzymes required for a key reaction.
   b. altering the shape of a key enzyme.
   c. becoming dormant when adequate quantities of the product are detected.
   d. destroying excess quantities of the product.
   e. storing enzymes until the product is exhausted.

6. A consequence of removing a phosphate from a molecule of ATP is the
   a. release of some energy.
   b. consumption of some energy.
   c. consumption of some oxygen.
   d. production of cellular waste.
   e. both a and d.

7. In plants, photosynthesis occurs in the
   a. central vacuole of most cells.
   b. cell wall.
   c. leaves only.
   d. chloroplasts.
   e. vascular tissue.

8. The actual formation of organic molecules from atmospheric carbon dioxide requires
   a. light.
   b. dark.
   c. ATP.
   d. NAD^+.
   e. NADH.

9. In a high school biology class, students observe that plants grown with a green filter in front of the light source grow better than plants with a red filter in front of the light source. Why is this?
   a. Plants do not absorb the red wavelengths of light for photosynthesis.
   b. Plants do not absorb the green wavelengths of light for photosynthesis.
   c. Plants reflect the red wavelengths of light during photosynthesis.
   d. Plants cannot reflect light during photosynthesis.
   e. Both a and b.

10. The presence of carotenoid pigments can be demonstrated by
    a. observing mature leaves during the hot summer months.
    b. bleaching leaves to remove all of the chlorophyll.
    c. observing leaves in the cool fall months.
    d. using a light microscope.
    e. both a and d.

11. Chlorophyll *a* and chlorophyll *b* absorb slightly different wavelengths of light primarily due to differences in
    a. the primary chlorophyll molecule.
    b. differences in "side groups" attached to the molecule.
    c. light quality.
    d. the carotenoid pigments available.
    e. none of the above.

12. Oxygen is produced during photosynthesis when
    a. the oxygen is removed from carbon dioxide to make carbohydrate.
    b. hydrogen is added to carbon dioxide to make carbohydrate.
    c. water molecules are split to provide electrons for photosystem I.
    d. water molecules are split to provide electrons for photosystem II.
    e. oxygen is added to chlorophyll.

13. Plants living in the desert conserve water at least in part by
    a. closing stomata when the temperature exceeds a critical point.
    b. increasing their uptake of water during the summer months.
    c. growing in groups with other related plants.
    d. absorbing increased amounts of oxygen in the summer.
    e. both a and c.

14. A consequence of increased oxygen concentration in the leaves of a plant are
    a. decreased absorption of light by chlorophyll.
    b. reduction in RuBP concentration in leaves.
    c. photorespiration in the leaves.
    d. more efficient photosynthesis.
    e. overproduction of carotenoid pigments.

15. Which of the following reactions is most likely to occur in the cytoplasm of a eukaryotic cell and in the absence of oxygen?
    a. chemiosmosis
    b. substrate-level phosphorylation
    c. fermentation
    d. glycolysis
    e. photosynthesis

Completion

1. Reactions that require energy are called __________, while reactions that release energy are __________ .
2. Enzymes that act as catalysts bind to the __________ of the substrate.
3. NAD^+ is an example of a(n) __________ .
4. Photosystem I absorbs light at __________ wavelength while photosystem II absorbs light at __________ wavelength.
5. The Calvin cycle takes place __________ and produces __________ .

Cool Places on the 'Net

You can find lecture notes and other materials related to an undergraduate biochemistry course at the World Lecture Hall. The address is:

http://www.utexas.edu/world/lecture/bch/

Label the Art

Label the parts of the mitochondrion.

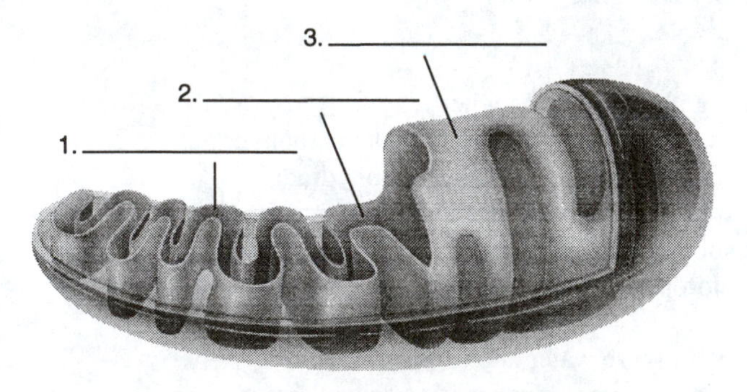

Crossword

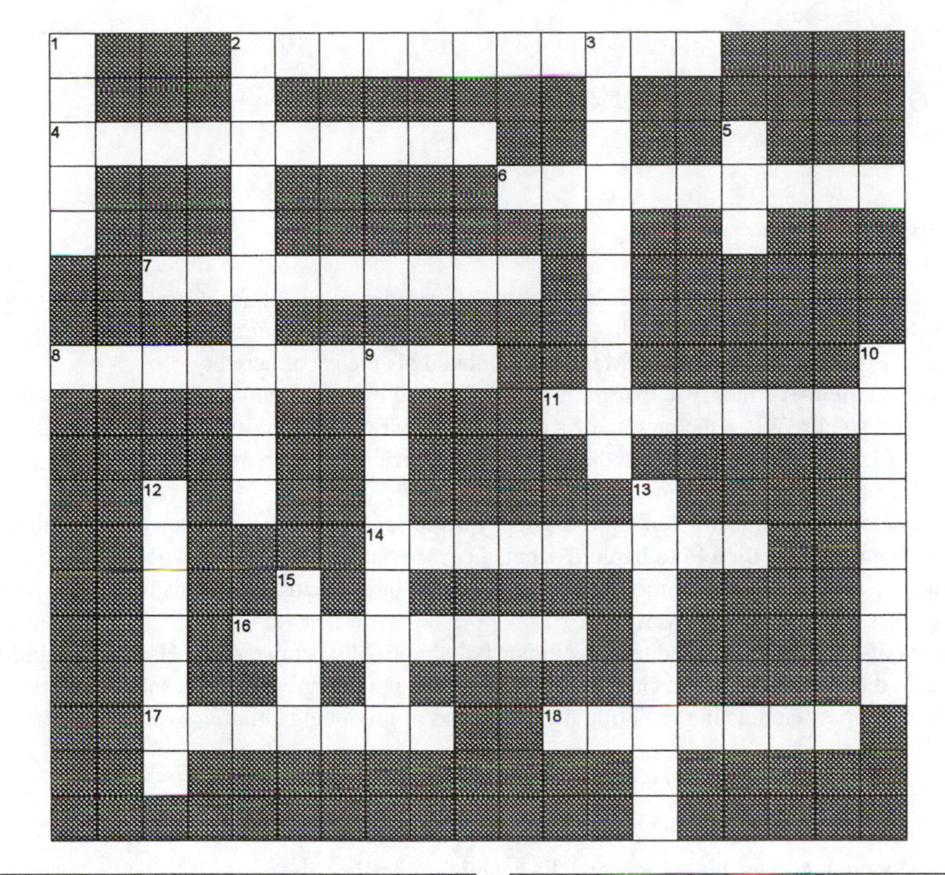

ACROSS

2. Photosynthesis organelle.
4. Reactions that require energy.
6. Takes place in mitochondria.
7. The harvesting of chemical energy from food is this type of reaction.
8. Does not require oxygen, takes place in the cytoplasm.
11. Nonprotein organic molecule.
14. A stack of grana
16. Enzyme that speeds up a chemical reaction.
17. Reactions that occur at the same time are called these.
18. Waxy covering on leaves.

DOWN

1. The color of leaves.
2. Photosynthetic pigments.
3. ________ energy is the energy required to initiate a chemical reaction
5. Energetic molecule.
9. The substance to be changed in a reaction.
10. A chemical ________ is the making or breaking of chemical bonds.
12. This is what a chemical reaction makes.
13. Some metal ions are these, they aid catalysis.
15. Electron carrier.

6 Foundations of Genetics

Chapter Concepts

In the 1860s, Gregor Mendel conducted genetic crosses of varieties of pea plants. Carefully counting the numbers of each kind of offspring, Mendel observed that in crosses of heterozygous parents, one-quarter of the offspring always appear recessive. From this simple result, Mendel formulated his theory of heredity.

The essence of Mendel's theory is that traits are determined by *information,* which was soon shown to be stored in genes on chromosomes. When the pattern of a trait's heredity reflects chromosome segregation, that trait is said to be "Mendelian." Many factors can obscure the underlying pattern of chromosomal segregation, such as when one gene modifies the phenotypic expression of another.

Haploid gametes are produced from diploid cells by a special form of cell division called meiosis. The reduction in chromosome number from diploid to haploid occurs because the chromosomes do not replicate between the two meiotic divisions. Early in meiosis, homologous chromosomes pair up closely. At this time, they often exchange segments, a process called crossing-over.

Extra copies of chromosomes produce developmental abnormalities or sterility. Harmful mutations that are inherited are called heredity disorders. Genetic counseling through pedigree analysis and techniques like amniocentesis are important aids in predicting the likelihood of producing children expressing hereditary disorders.

Multiple Choice

1. Garden peas were a good subject for Mendel's genetic studies because
   a. they are available in many varieties.
   b. they are small, easy to grow plants.
   c. it is easy to control their fertilization.
   d. they mature quickly.
   e. all of the above.

2. To conduct genetic experiments it is necessary to begin with subjects that are
   a. haploid.
   b. heterozygous.
   c. true-breeding.
   d. easy to count.
   e. previously unstudied.

3. During your plant research project, you cross a blue-flowered variety with a yellow-flowered variety. The F_1 generation consists of all yellow-flowered plants. The yellow-flowered plants are probably
   a. recessive.
   b. dominant.
   c. incompletely dominant.
   d. sterile.
   e. hybrids.

4. In the experiment mentioned above the purple-flowered plants are probably
   a. recessive.
   b. dominant.
   c. incompletely dominant.
   d. sterile.
   e. hybrids.

5. In the experiment mentioned above what would you expect to see in the F_2 generation?
   a. all yellow-flowered plants
   b. all purple-flowered plants
   c. yellow-flowered and purple-flowered plants
   d. plants with an intermediate color between yellow and purple
   e. none of the above.

6. Which of the following represents an individual that is homozygous for the dominant trait?
   a. *bb*
   b. *Bb*
   c. *BB*
   d. *aa*
   e. *Aa*

7. Which of the following represents an individual that is heterozygous for the dominant trait?
   a. *bb*
   b. *Bb*
   c. *BB*
   d. *aa*
   e. *AA*

8. Two common coat colors seen in Shetland sheepdogs (shelties) are sable and tricolor. Sable dogs (*B*) are blonde and tricolored dogs (*b*) are black with a white collar and tan eyebrows. Sable is the dominant color; tricolor is recessive. A dog that is heterozygous for coat color (*Bb*) is called a tri-factored sable and is darker than a dog that is homozygous for sable (*BB*). What are the possible coat colors in a litter that results from a cross between a homozygous sable and a tri-factored sable?
   a. all sable puppies
   b. all tricolored puppies
   c. both sable and tricolored puppies
   d. sable and tri-factored sable puppies
   e. all tri-factored sable puppies

9. What coat colors are possible from a cross between a sable and a tricolored dog?
   a. all sable puppies
   b. all tricolored puppies
   c. both sable and tricolored puppies
   d. sable and tri-factored sable puppies
   e. all tri-factored sable puppies

10. A tri-factored sheltie has a litter of puppies. Two of the pups are tricolored and the other two are tri-factored. What color was the father of the puppies?
    a. sable
    b. tricolored
    c. tri-factored
    d. either a or b
    e. either b or c

11. If the tri-factored sheltie had one sable puppy and three tri-factored puppies, then the father would have been what color?
    a. sable
    b. tricolored
    c. tri-factored
    d. either a or b
    e. either b or c

12. A cross between two individuals results in a ratio of 9:3:3:1 of four possible phenotypes. This is an example of a(n)
    a. monohybrid cross.
    b. dihybrid cross.
    c. testcross.
    d. incomplete dominance.
    e. none of the above.

13. When a trait is controlled by the interaction of the products of two genes it is referred to as
    a. dominance.
    b. pleiotropy.
    c. epistasis.
    d. incomplete dominance.
    e. aneuploidy.

14. The tri-factored shelties mentioned earlier are an example of
   a. dominance.
   b. pleiotropy.
   c. epistasis.
   d. incomplete dominance.
   e. aneuploidy.

15. A DNA molecule that is complexed with proteins into a rodlike structure is called a(n)
   a. chromophore.
   b. chromosome.
   c. genophore.
   d. genome.
   e. genosome.

16. The end result of meiosis in human males is
   a. two diploid sperm cells.
   b. four diploid sperm cells.
   c. two haploid sperm cells.
   d. four haploid sperm cells.
   e. none of the above.

17. __________ is a significant source of genetic recombination during gamete production.
   a. Mutation
   b. Crossing-over
   c. Nondisjunction
   d. Controlled assortment
   e. Both a and c

18. Which of the following is a result of nondisjunction of chromosomes during gamete formation?
   a. cystic fibrosis
   b. Down syndrome
   c. Turner syndrome
   d. sickle-cell anemia
   e. Both b and c

19. A Barr body is a(n)
   a. inactivated X chromosome.
   b. inactivated Y chromosome.
   c. an abnormality seen in the cells of persons with Down syndrome.
   d. chromosome 21 trisomy complex.
   e. none of the above.

Completion

1. Hereditary factors are called __________ , while alternate forms of those factors are called __________ .
2. The possible genotypes that can be achieved by a cross of *AB* and *ab* are __________ .
3. The frequency of genotypes that occur in the offspring of a cross is usually expressed by a(n) __________ .
4. A(n) __________ could be used to determine if an individual has a *WW* or a *Ww* genotype.
5. An allele that controls more than one trait is said to be __________ .
6. Sperm cells have a __________ number of chromosomes, while skin cells have a __________ number of chromosomes.
7. A picture of the chromosomes contained in a cell is called a(n) __________ .

Cool Places on the 'Net

To search for information on the latest research in human genetics check out the American Society of Human Genetics home page at:

http://www.faseb.org/genetics/ashg/ashgmenu.htm

Label the Art

Complete the following Punnett squares.

A.

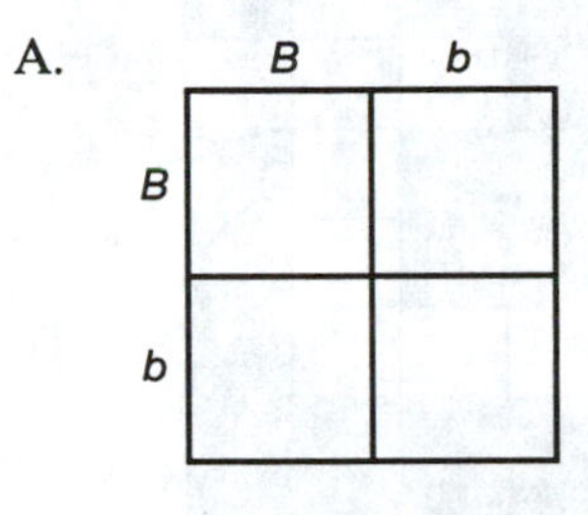

B.

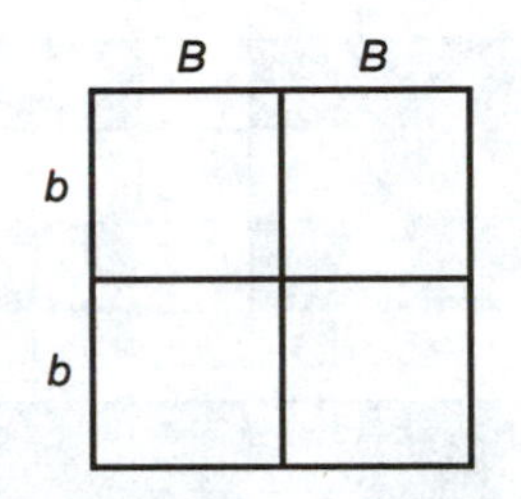

C.

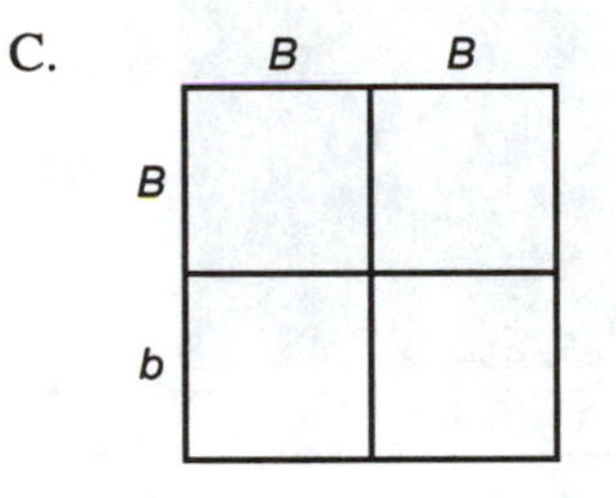

D.

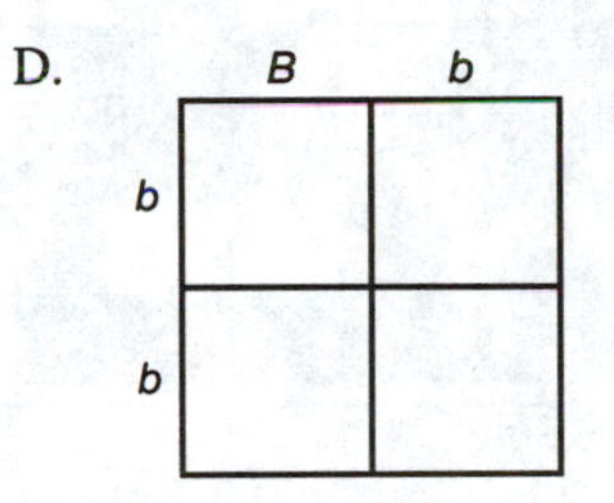

Crossword

ACROSS	
2.	A result of nondisjunction.
4.	The _____ of segregation.
5.	One allele affects more than one trait.
6.	Has two copies of the same gene.
7.	A picture of chromosomes.
11.	The garden _____ was an early subject for genetics studies.
13.	A trait that is expressed.
15.	One gene modifies the phenotypic expression of another.
17.	_______ syndrome: XO gneotype.
19.	Traits passes from parents to offspring.
20.	Trait that is not expressed unless present as homozygous.

DOWN	
1.	Different forms of the same gene.
3.	______ dominance: heterozygote resembles one allele but can be distinguished from it.
8.	Your genetic makeup is this.
9.	Genetic information is in this form.
10.	______ square is used to determine possible results of a test cross.
12.	How you appear is this.
14.	Two copies of each gene.
16.	Alternate form of a gene.
18.	The father of genetics.

7 How Genes Work

Chapter Concepts

A series of critical experiments demonstrated that genes are composed of DNA. A key advance was the demonstration by Avery that the active ingredient in Griffith's transforming extract was DNA. DNA is a double helix, in which A on one strand pairs with T on the other, and G similarly pairs with C.

DNA serves as a template on which mRNA is assembled in a process called transcription. The genetic information is encoded in DNA in three-nucleotide units called codons. The genetic code specifies which codons correspond to each amino acid. The beginning of a gene, where the RNA polymerase binds the DNA, is the promoter.

mRNA serves as the template guiding assembly of proteins. A ribosome moves along the mRNA, adding an amino acid to the end of a growing protein chain as it passes each codon.

Most genes are regulated by controlling the rate at which they are transcribed. Bacteria shut off genes by attaching a repressor protein to or near the promoter, so the polymerase is unable to bind. Eukaryotes can use many regulatory proteins simultaneously.

Multiple Choice

1. The material that transformed harmless bacteria into virulent bacteria in Avery's experiment was
   a. DNA.
   b. mRNA.
   c. tRNA.
   d. protein.
   e. RNA polymerase.

2. What is the significance of Chargaff's rule in the DNA molecule?
   a. The pairing of purines and pyrimidines in DNA results in an excess of purine in some molecules.
   b. Chargaff's rule says that a double helix should not form except at pH 7.
   c. The rules of base pairing proposed by Chargaff result in a molecule with a constant thickness.
   d. Both a and b.
   e. None of the above.

3. The replication of DNA molecules in a eukaryotic cell takes place
   a. just prior to metaphase in mitosis.
   b. prior to cell division.
   c. continuously.
   d. only during meiosis.
   e. shortly after telophase.

4. What is the base sequence of a DNA strand that reads CCATAGTTTCA?
   a. CCATAGTTTCA
   b. GGATAGCCCAT
   c. GGTATCAAAGT
   d. GGTATCCGCAT
   e. none of the above

5. What would be the base sequence of an mRNA molecule that was complimentary to that DNA strand?
   a. GGTATCAAAGT
   b. CCATAGTTTCA
   c. CCUAUGAAAGU
   d. GGUAUACAAGU
   e. both a and b

6. The production of mRNA is called
   a. transduction.
   b. transformation.
   c. transcription.
   d. translation.
   e. promotion.

7. Which of the following is the most accurate description of RNA polymerase?
   a. RNA polymerase is used to initiate the production of tRNA from an mRNA template.
   b. RNA polymerase binds to a strand of DNA and moves along the strand to make a strand of mRNA.
   c. RNA polymerase is responsible for providing a termination signal to the DNA strand at the end of rRNA synthesis.
   d. DNA ligase and RNA polymerase work together to bind the nucleotides in a newly forming DNA strand.
   e. RNA polymerase joins the two DNA after translation has occurred.

8. During transcription
   a. both DNA strands serve as a template for the synthesis of mRNA.
   b. the two DNA strands alternate as the template for the synthesis of tRNA.
   c. one of the DNA strands serves as a template for the synthesis of mRNA.
   d. mRNA serves as a template for the synthesis of rRNA.
   e. rRNA is assembled into complexes with proteins to make new ribosomes.

9. Each amino acid is specified by
   a. several genes.
   b. an operon and a promoter.
   c. a codon.
   d. markers of the surface of the cell requiring it for growth.
   e. an enhancer.

10. In ribosomes the rRNA is located
    a. on the surface of the large subunit.
    b. complexed with several proteins in the large subunit.
    c. in both the small and the large subunits.
    d. in the small subunit.
    e. none of the above.

11. Why can ribosomes bind to a sequence occurring at the beginning of a gene?
    a. Because of the matching, exposed sequence of rRNA nucleotides found on the small subunit of the ribosome.
    b. Because of the effects of certain hormones that are associated with DNA synthesis.
    c. Because the conformation of the ribosome contributes to binding with DNA.
    d. Both a and b.
    e. None of the above.

12. What part of the tRNA molecule binds to mRNA?
    a. The single-stranded nucleotide "tail" of the tRNA molecule binds to a specific codon on the mRNA molecule.
    b. The anticodon loop on the tRNA molecule binds to a specific codon on the mRNA molecule.
    c. The tRNA codon binds to the anticodon on the mRNA molecule.
    d. The small subunit of the tRNA binds to the promoter on the mRNA molecule.
    e. The large subunit of the tRNA molecule binds to the anticodon on the mRNA molecule.

13. What happens when a ribosome reaches a codon on the mRNA molecule that does not specify any of the 64 tRNA anticodons?
    a. The ribosome begins to read the mRNA molecule in the opposite direction.
    b. The ribosome releases the newly formed protein molecule.
    c. The ribosome subunits separate.
    d. The newly formed protein begins to break up into smaller molecules.
    e. Both b and c.

14. Which of the following mechanisms would function to prevent transcription?
    a. the addition of excess substrate to the environment
    b. binding of a repressor to the operator site on the gene
    c. binding of an enhancer to the repressor site on the gene
    d. changing the pH of the environment
    e. both a and d

15. What effect would the addition of lactose have on a repressed *lac* operon?
    a. The operator site on the operon would move.
    b. It would reinforce the repression of that gene.
    c. It would result in the repression of other genes in that area.
    d. The *lac* operon would be transcribed.
    e. It would have no effect whatsoever.

16. After transcription, the mRNA molecule must __________ before it is translated into a protein molecule.
    a. assume the correct three-dimensional sequence
    b. adjust to its new environment
    c. have its introns removed
    d. have its exons removed
    e. complex with a tRNA molecule

17. Chemical or physical damage to DNA can cause __________ mutations involving one or a few nucleotides.
    a. single
    b. simple
    c. insignificant
    d. point
    e. patterned

Completion

1. The DNA parent molecule begins to unzip at the __________ .
2. The conservation of one parental strand of DNA when it is copied is called __________ replication.
3. A cluster of genes that are transcribed as a unit are called a(n) __________ .
4. Coding sequences of nucleotides are called __________ , while noncoding sequences are __________ .
5. Substances that are capable of causing damage to DNA are called __________ .

Cool Places on the 'Net

To find out about a variety of health-related issues and to read about the latest research check out the *New England Journal of Medicine* OnLine at:

http://www.nejm.org/JHome.htm

Label the Art

Fill in the five steps of protein synthesis on the following figure.

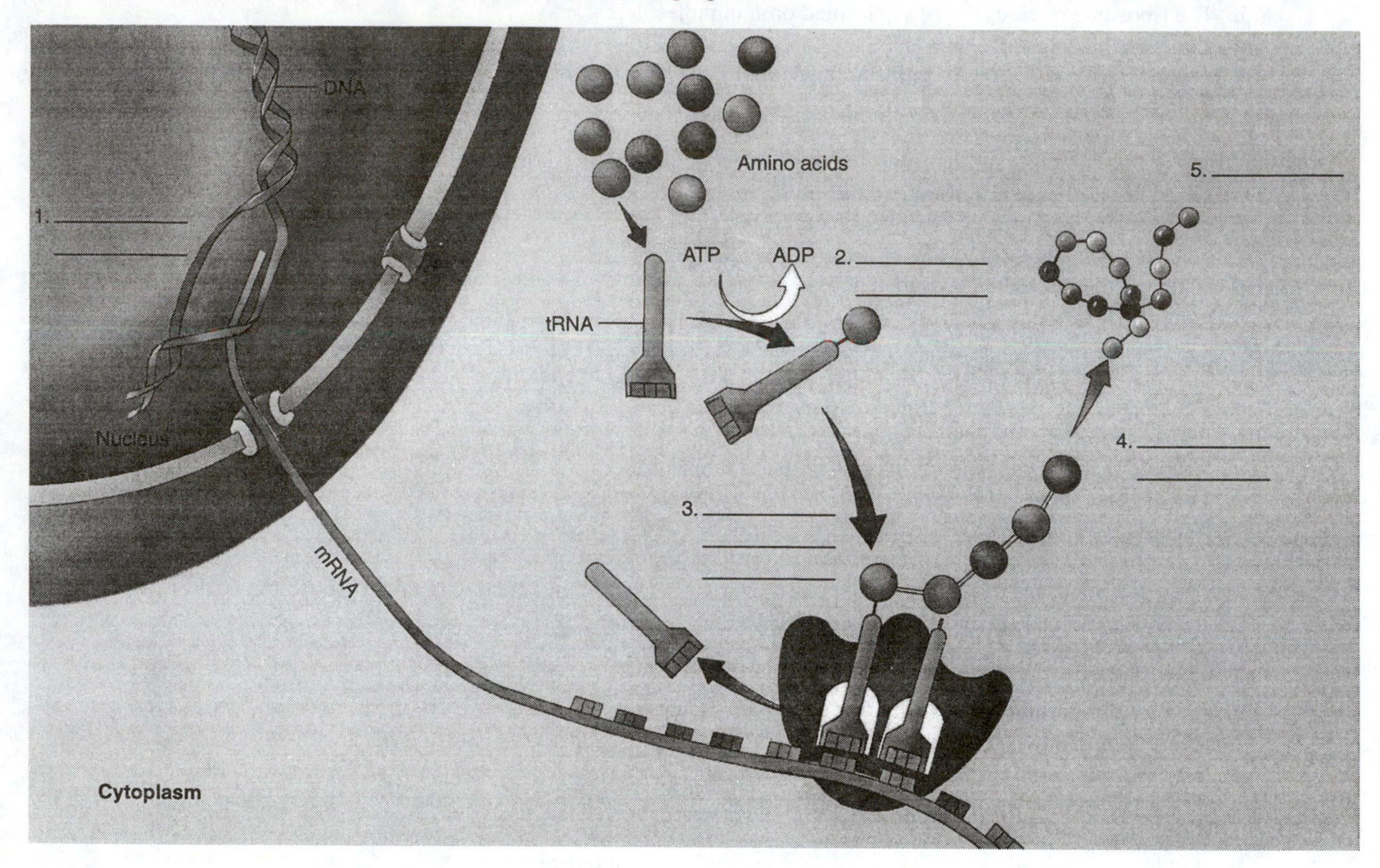

Crossword

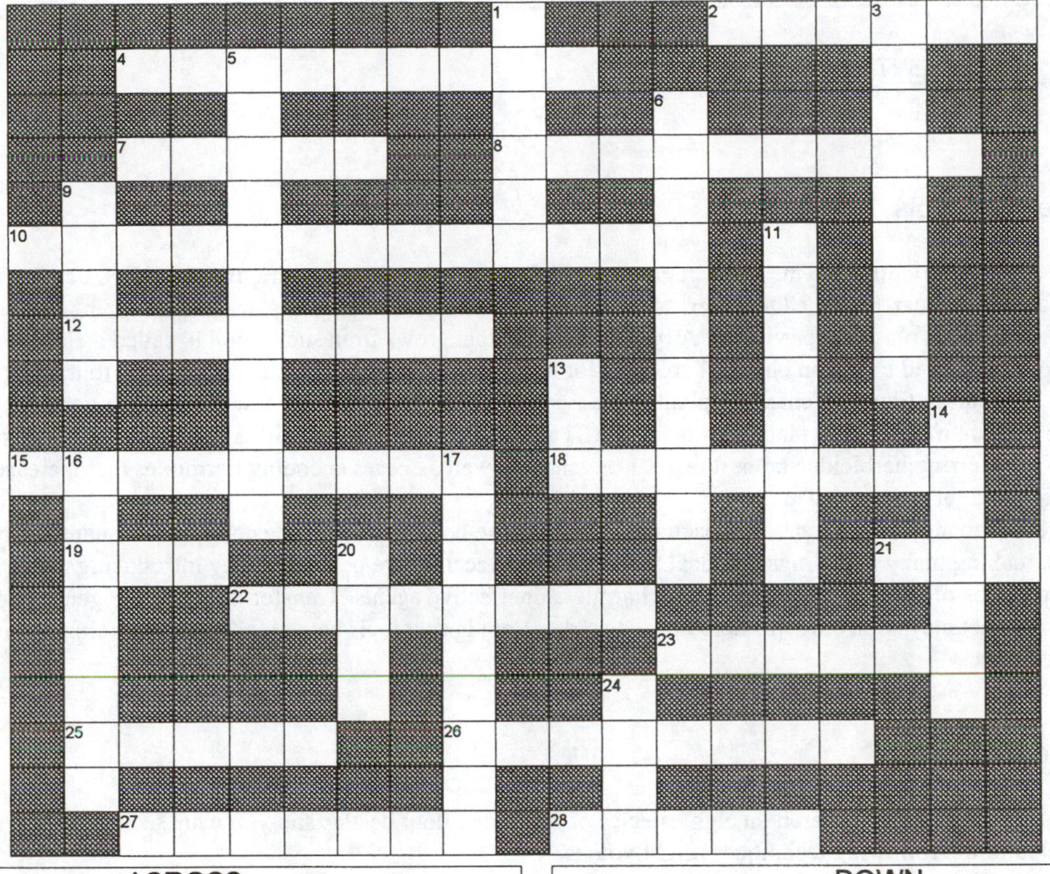

ACROSS

2. A segment of DNA containing a cluster of genes that are transcribed as a unit.
4. This is where tRNA binds.
7. Worked with viruses that infect bacteria.
8. This is what blocks the movement of polymerase.
10. The first step in gene expression.
12. Pyrimidine that pairs with guanine.
15. The "m" in mRNA means this.
18. He worked with Avery.
19. Short for Holmes ribgrass virus.
21. Short for tobacco mosaic virus.
22. Radioactive ________ are used to label DNA.
23. American scientist that worked on the structure of DNA.
25. Protein-coding regions of DNA.
26. A change in a cell's genetic message.
27. "Extra stuff" in DNA.
28. DNA is a double ________ molecule.

DOWN

1. Worked with tobacco mosaic virus.
3. The "r" in rRNA means this.
5. DNA from disease causing strains of bacteria moved to harmless strains by this process.
6. mRNA directs production of a protein.
9. English scientist that worked on determining the structure of DNA.
11. The "t" in tRNA means this.
13. Radioactive labels are used to _____ and track proteins.
14. Pyrimidine that pairs with adenine.
16. Bind signal proteins.
17. The site of protein synthesis.
20. DNA begins to unwind at a replication _______.
24. ______ pairing of nucleotides.

8 Gene Technology

Chapter Concepts

DNA is cut by restriction enzymes into fragments with "sticky" ends. Any two fragments of DNA cut by the same restriction enzyme can be joined together, whatever the source of the DNA. DNA fragments can be introduced into bacteria, carried on plasmids or viruses. A bacterial colony that grows from such a cell is called a clone. Probes such as cDNA can be used to screen bacterial colonies for the presence of a gene you are attempting to transfer.

It is difficult to transfer genes into plants because of their tough cell walls—few viruses can penetrate them. Today genes are inserted into plant cells using the Ti plasmid or they are shot with a gun. Major advances have been made in transferring herbicide and pest resistance genes, as well as genes encoding hormones that increase milk production and farm animal size.

By transferring human genes into bacteria, it has become possible to produce commercial quantities of rare proteins, making many new drugs practical. Effective new vaccines are being made by introducing genes specifying surface proteins of pathogenic microbes into harmless, ineffective agents. Transfer of healthy *cf* genes into cystic fibrosis patients may soon cure this hereditary disorder. Rapid progress is being made in sequencing the entire human genome.

Multiple Choice

1. You are conducting research on eight species of *Tribolium* flour beetles and you want to compare their proteins. Which of the following techniques might you use?
   a. genetic engineering
   b. gene amplification
   c. gel electrophoresis
   d. polymerase chain reaction
   e. gene therapy

2. The key to isolating a particular gene for use in recombinant DNA research is to
   a. find a restriction enzyme to cut out the gene.
   b. find a compatible species to receive the gene.
   c. identify copies of the gene on several chromosomes.
   d. both a and b.
   e. none of the above.

3. Cuts in DNA are sealed with
   a. restriction enzymes.
   b. ligases.
   c. reverse transcriptase.
   d. polymerase.
   e. plasmids.

4. DNA is made from RNA with the help of
   a. restriction enzymes.
   b. ligases.
   c. reverse transcriptase.
   d. polymerase.
   e. plasmids.

5. Sticky ends are created by
   a. adding ligase to the end of the DNA fragment.
   b. adding polymerase to the end of the DNA fragment.
   c. cutting DNA so that there are short, single strands at each end.
   d. cutting the DNA fragment in several places.
   e. both a and d.

6. When investigating a crime scene detectives find a single hair on the floor. The sample is not large enough to perform many tests. Which of the following methods could be used to make sufficient DNA to allow for comparison to the DNA of the prime suspect?
   a. Reverse transcriptase could be used to make DNA from RNA in the hair sample.
   b. The hair root could be tissue cultured to make additional copies of the DNA.
   c. The gene could be amplified by use of electrophoresis.
   d. Polymerase chain reaction could be used to make many copies of the DNA.
   e. The gene of interest could be spliced into a bacterium that would make a new copy every time it replicated.

7. cDNA is significant when
   a. you need several thousand copies of a gene.
   b. cloning eukaryotic DNA in prokaryotic cells.
   c. it is necessary to copy introns in prokaryotic cells.
   d. a primary transcript is not available.
   e. processed mRNA is not available.

8. In order for plants to resist the effects of the herbicide Roundup, they must be
   a. monocots.
   b. dicots.
   c. annuals.
   d. able to synthesize glyphosate.
   e. resistant to glyphosate.

9. Although the T_i plasmid has revolutionized plant genetic engineering, one limitation of its use is that it
   a. is only about 50% effective.
   b. can only be used on fruit-bearing plants.
   c. is only effective on plants that are resistant to Roundup.
   d. does not infect cereal plants.
   e. cannot transmit prokaryotic genes.

Matching

1. _____ Bovine growth hormone
2. _____ Factor VIII
3. _____ Human genome
4. _____ Electrophoresis
5. _____ Insulin
6. _____ Nitrogen fixation
7. _____ Probe
8. _____ Glycophosphate

a. Related to the ability to control blood sugar levels
b. Increases soil fertility
c. A herbicide and the active ingredient in Roundup
d. Increased milk production in dairy cows
e. Promotes blood clotting
f. Mapping all human genes
g. Used to detect the presence of genes of similar sequence
h. A procedure that separates substances based on their size and electrical charge

Completion

1. Implanting a human gene for insulin into a bacterium is an example of __________ .
2. DNA cut with restriction enzymes have __________ at each end of the fragment.
3. List the stages involved in the transfer of genes from one species to another using genetic engineering: __________ , __________ , __________ , and __________ .
4. A(n) __________ is used to detect the presence of genes with a similar sequence.
5. __________ is the ability of some bacteria to convert atmospheric nitrogen into a form that plants can use. This is important because it __________ soil fertility.

Cool Places on the 'Net

You can find information on biotechnical research and genetic engineering at the Agricultural Biotechnology Center. Their address is:

http://www.abc.hu/

Bio Online provides access to information related to biotechnical and pharmaceutical research. Their address is:

http://www.bio.com/

Key Word Search

C P K W S C Q P Y E A T E G O
Y L I G A S E M Y Z N E L S Y
S I S E R O H P O R T C E L E
T P I N S U L I N V I T G H A
I B D E I A N R P P C R U E W
C C G T W N R I Q L O M G S J
F S W H C R Z E Y B A Z K K S
I R A E Z M B O M N G S X Z L
B S D R R O V D G Y U C M T J
R R O A R L W E K H L T T I Z
O A J P K M N D Z P A O J W D
S E G Y Q O L S Z G N Z P T V
I O V H M A I T G P T D M R A
S B T E Y R Y M F H S P N F U
D G J T G T P Q W I O L M T Z

Anticoagulants
BGH
Cystic fibrosis
Electrophoresis
Enzymes
Gene Therapy
Human genome
Insulin
Ligase
MRNA
PCR
Plasmid
Polymerase
Probe

9 Evolution and Natural Selection

Chapter Concepts

In 1831 Darwin sailed around the world, closely observing the plants and animals he saw. In 1859 Darwin published *On the Origin of Species,* in which he proposed that the mechanism of evolution was natural selection.

If one dates fossils, and orders them by age, progressive changes are seen. This is direct evidence that evolution has occurred. The progressive accumulation of molecular differences and comparisons of living organisms provide additional strong evidence that evolution has occurred.

In a population not undergoing significant evolutionary change, two alleles present in frequencies p and q will be distributed among the genotypes in the proportions $p^2 + 2pq + q^2$, the Hardy-Weinberg equilibrium. Allele frequencies change in nature due to mutation, migration, drift, nonrandom mating, and selection.

A mutation in hemoglobin causes a condition known as sickle-cell anemia. This recessive mutation is common in central Africa because it renders heterozygous individuals resistant to malaria. Vegetation darkened by industrial soot has favored the evolution of darker moths, better concealed from their predators.

Microevolution leads to macroevolution. Adaptation to local habitats leads to divergence in the evolution of ecological races. Isolating mechanisms then reinforce the differences, leading to reproductive isolation and species formation.

Multiple Choice

1. Darwin developed his theories on evolution by natural selection after observing
   a. the feeding adaptations of the Galápagos finches.
   b. fossils of extinct and living armadillos in South America.
   c. similarities and differences in the Galápagos finches and South American finches.
   d. differences in the reproductive success in a variety of organisms.
   e. all of the above.

2. A key point made by Thomas Malthus in his *Essay on the Principles of Population* was that
   a. natural selection is the mechanism of evolution.
   b. natural selection is more significant in wild animals than in domestic animals.
   c. the carrying capacity of any population is limited by the available natural resources.
   d. populations of species remain more or less stable in size.
   e. reproductive success is high in live-bearing species.

3. A paleontologist who is dating a fossil using the carbon-14 method is measuring the
   a. electron decay rate of the isotope.
   b. length of time it takes for carbon-14 to obtain another proton in its nucleus.
   c. half-life of the isotope.
   d. time it takes for the isotope to decay to hydrogen.
   e. none of the above.

4. Evolution
   a. has occurred only in the past 150 million years.
   b. involves a progression of changes that took place over long periods of time.
   c. occurs more often in plant species than in animals.
   d. is no longer possible.
   e. always results in species adapted for the current environment.

5. Molecular clocks measure
   a. differences in nucleotide sequences between different species.
   b. the age of fossils more than 150 million years old.
   c. the half-life of potassium-16.
   d. both a and b.
   e. both b and c.

6. After examining the evidence related to the evolution of hemoglobin you might conclude that
   a. lamprey globin evolved prior to insect globin.
   b. plant globins evolved from an entirely different type of gene than did human hemoglobin.
   c. it is strong evidence that modern organisms evolved from simpler forms.
   d. fish are more closely related to whales than they are to kangaroos.
   e. both a and d.

7. Insect wings and the wings of bats are
   a. analogous structures.
   b. homologous structures.
   c. simple structures.
   d. structurally similar.
   e. the result of microevolution.

8. In the year 2317, a group of 507 pioneers colonize Mars. The incredible distance between Mars and Earth excludes the possibility of new members joining the colony. This situation is an example of
   a. genetic drift.
   b. mutation.
   c. the founder effect.
   d. artificial selection.
   e. outcrossing.

9. Which of the following conditions are necessary for the Hardy-Weinberg theorem to be valid?
   a. The population being examined must be small and have a limited gene pool.
   b. New alleles must be added to the population on a regular basis.
   c. All of the individuals in the population must be heterozygous for all traits.
   d. Natural selection is acting on the population.
   e. Random mating is occurring within the population.

10. Much of the soil in Sedona, Arizona, is brick red in color. Several birds and small mammals living in that area include ants in their diet. Although both black and red ants live there, red ants are much more common. This might be due to
    a. directional selection.
    b. stabilizing selection.
    c. disruptive selection.
    d. outcrossing.
    e. heterozygote advantage.

11. Two species of birds that do not mate because they respond to different mating dances are separated by
    a. geographical isolation.
    b. a prezygotic isolating mechanism.
    c. behavioral isolation.
    d. a postzygotic isolating mechanism.
    e. both b and c.

Matching—Part 1

1. _____ Kettlewell
2. _____ Wallace
3. _____ Darwin
4. _____ Hardy-Weinberg
5. _____ Malthus

a. Developed the theory of evolution by natural selection
b. Wrote essay on the "Principles of Population"
c. Studied industrial melanism
d. Wrote a paper on evolution but did not provide as much evidence as was in *On the Origin of Species*
e. Studied how changes in gene frequencies lead to evolution

Matching—Part 2

1. _____ Mutation
2. _____ Migration
3. _____ Genetic drift
4. _____ Founder effect
5. _____ Bottleneck effect
6. _____ Artificial selection
7. _____ Directional selection
8. _____ Stabilizing selection
9. _____ Disruptive selection

a. The intermediate type phenotype is eliminated from a population
b. When pollution darkened tree bark in England the light-colored moths declined in numbers
c. The drifting of gametes in plants is an example of this
d. A decrease in the genetic variability in cheetahs has caused this to happen
e. Due to an error in the replication of a nucleotide sequence in DNA
f. Breeding plants and animals with special characteristics
g. The random loss of alleles from a small population
h. Both phenotypic extremes are eliminated from a population
i. A few finches colonizing the Galápagos Islands

Completion

1. Fossils of __________ , a link between reptiles and birds, are about _________ years old.
2. Dating fossils by measuring carbon-14 decay is called _________ and can be used to date fossils __________ years old.
3. A high number of homozygous individuals in a population can be the effect of __________ .
4. __________ results in the increased frequency of individuals with the intermediate phenotype in a population.
5. Organisms that have the potential to interbreed in nature are members of the same __________ .

Cool Places on the 'Net

To find out more about evolution and to explore interesting ideas such as evolution versus creationism check out:

http://www.yahoo.com/Science/Biology/Evolution

Key Word Search

```
M T F I R D C I T E N E G X G
O C S D M A S R M M S I P R M
L E T I I L P Y A B L Z E I R
E F A S G L E V C I A B C M A
C F B R R E C A R W N R N S D
U E I U A L I H O I O H O I I
L R L P T E E W E E I H I L O
A E I T I S S W V F T N T A I
R D Z I O S Y O O D C A A U S
C N I V N D L S L H E R L D O
L U N E R U S P U R R D U A T
O O G A T I V Q T T I D P R O
C F H I L T E X I K D H O G P
K P O S T Z Y G O T I C P G E
G N I D E E R B N I F A M T W
```

- Alleles
- Directional
- Disruptive
- Fossils
- Founder effect
- Genetic drift
- Gradualism
- Hardy Weinberg
- Inbreeding
- Macroevolution
- Microevolution
- Migration
- Molecular clock
- Population
- Postzygotic
- Radioisotope
- Species
- Stabilizing
- Stabilizing

Crossword

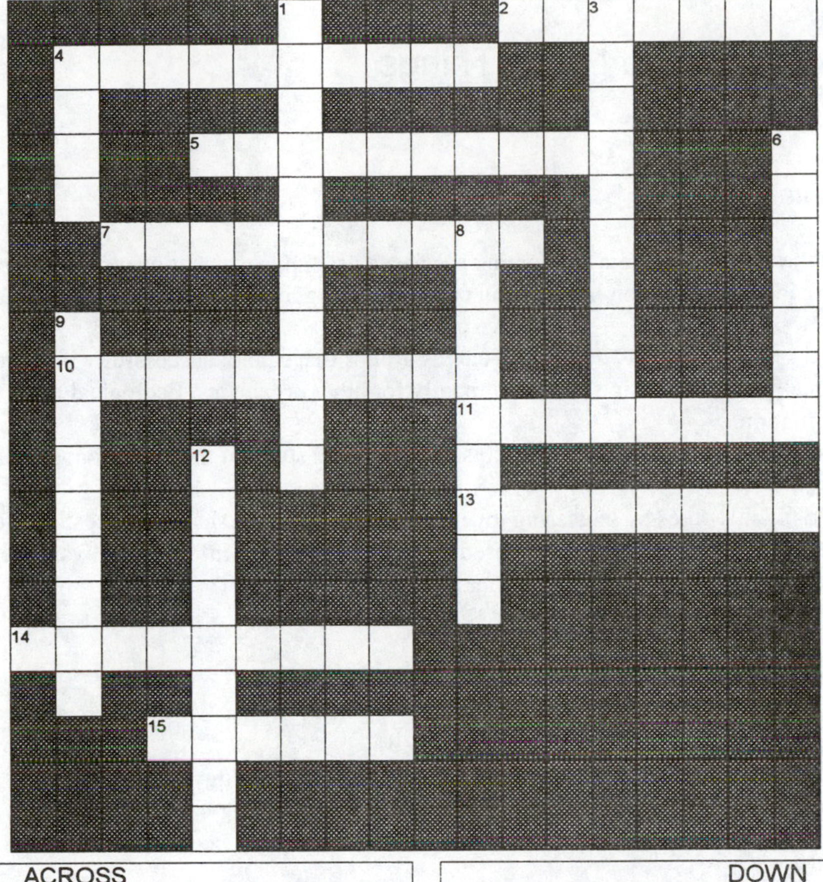

ACROSS	DOWN

ACROSS

2. On the Origin of _______ is a book that shook things up in the 1800s
4. _______ structures have similar evolutionary origins
5. Over 300 breeds of dogs have been produced using this type of selection
7. Mechanisms that prevent fertilization between two species
10. This type of melanism was seen in post-Industrial Revolution moths in England
11. _______ structures that evolved for similar function, such as the wings of insects and birds
13. Species are reproductively _______
14. A _______ clock measures relatedness of species based on differences in amino acid sequences
15. _______ cell anemia

DOWN

1. Mechanisms that result in no offspring or sterile offspring
3. These races are much slower than foot races
4. Radioactive decay is measured as _____-life
6. Petrified organisms
8. Species are kept separate by various _______ mechanisms
9. The movement of genetic information into and out of a population
12. Natural _______ means survival of the most fit

10 How We Name Living Things

Chapter Concepts

Linnaeus invented the binomial system for naming species, a vast improvement over complex polynomial names. Linnaeus placed organisms into seven hierarchical categories: kingdom, phylum, class, order, family, genus, and species.

The "biological species concept" defines species as groups that cannot successfully interbreed. This concept works well for animals and outcrossing plants but poorly for other organisms. Ecological races are an intermediate stage in species formation.

Taxonomy is of great practical importance. Systematics is the study of the evolutionary relationships among a group of organisms.

Cladistics builds family trees by clustering together those groups that share an ancestral "derived" character. Phylogenetic systematics classifies organisms based on a large amount of information available, giving due weight to the evolutionary significance of certain characters.

Multiple Choice

1. Arranging organisms into a multilevel system is called
   a. classification.
   b. nomenclature.
   c. speciation.
   d. paleontology.
   e. ornithology.

2. In the name *Escherichia coli, coli* is the __________ name.
   a. genus
   b. species
   c. family
   d. common
   e. phylum

3. The difficulty with polynomial names was that they
   a. were not descriptive enough.
   b. were not consistent.
   c. could only be applied to animals.
   d. were long and confusing.
   e. all of the above.

4. Taxonomy
   a. is closely related to classification.
   b. identifies and names taxa of organisms.
   c. names specific organisms.
   d. both a and b.
   e. both b and c.

5. Which of the following is the correct way to write a scientific name?
   a. Canis familiaris
   b. Canis Familiaris
   c. canis familiaris
   d. <u>Canis familiaris</u>
   e. *Canis familiaris*

6. Which of the following is the correct order of taxonomic categories beginning with the largest and ending with the smallest group?
   a. phylum, kingdom, class, family, genus, order, species
   b. kingdom, phylum, class, order, family, genus, species
   c. class, family, species, order, phylum, kingdom, genus
   d. species, genus, family, order, class, phylum, kingdom
   e. none of the above

7. Organisms that are members of the same class are also members of the same
   a. species.
   b. family.
   c. genus.
   d. phylum.
   e. both b and d.

8. While examining water samples that you collected at a local pond, you observe a unicellular organism that has a nucleus and organelles. That organism is a member of the kingdom
   a. Eubacteria.
   b. Protista.
   c. Animalia.
   d. Plantae.
   e. Fungi.

9. One limitation of the biological species concept is that it
   a. does not apply to asexually reproducing organisms.
   b. is only applicable to plant species.
   c. is only applicable to animal species.
   d. is outdated.
   e. all of the above.

10. While conducting research in the rain forest you collect a number of beetles. Several of them have similar morphologies but different coloring. How could you determine if they are members of the same species?
    a. by determining if they produce viable, fertile offspring
    b. by determining if they live in the same niche
    c. by determining if they eat the same foods
    d. by further examination of their morphologies for significant differences
    e. both a and d.

11. It is later determined that the beetles mentioned above are able to produce viable, fertile offspring. The different marking could mean that they are
    a. morphologically isolated.
    b. ecotypes.
    c. ecological races.
    d. serovars.
    e. both b and c.

12. Phylogeny
    a. refers to the evolutionary relationships between organisms.
    b. provides information about community interactions.
    c. explains the physiology of organisms.
    d. is only valid in relation to sexually reproducing organisms.
    e. both a and b.

13. Which of the following is a difference between cladistics and phylogenetic systematics?
    a. Phylogenetic systematics reflects relative relationships between organisms.
    b. Cladistics relies more heavily on morphological characteristics than physiological characteristics.
    c. The cladistic approach provides a more accurate picture of the evolutionary relationships between organisms.
    d. Phylogenetic systematics is the better approach when information about the organisms being studied is plentiful.
    e. both a and b.

14. Although dogs and coyotes can produce fertile offspring, they are not members of the same species. Why is this?
    a. differences in morphology
    b. differences in their behaviors
    c. many of their offspring are sterile
    d. they inhabit different environments
    e. they require different amino acids in their diets

15. Although two birds living in the same forest share many similarities, they rarely interbreed. One species lives in the upper branches of trees while the other lives in low-growing shrubs. Their different niches act as a(n)
    a. separating mechanism.
    b. isolating mechanism.
    c. species isolator.
    d. cladistic barrier.
    e. none of the above.

Completion

1. The taxonomic group between order and genus is __________ .
2. Bacteria and protists reproduce __________ .
3. When using cladistics to explore the phylogeny of a group of organisms, it is necessary to an organism to serve as a baseline for comparison. This organism is referred to as the __________ .
4. Approximately __________ species have been named since Linnaeus developed the binomial system of nomenclature.

Cool Places on the 'Net

For the latest word on the classification of organisms, visit the home page of the Classification Society of North America. This organization promotes scientific study of classification and disseminates scientific and educational information. They can be found at:

http://www.pitt.edu/-csna

Key Word Search

M	E	C	O	T	Y	P	E	S	T	B	B	I	M	W
P	E	C	O	L	O	G	I	C	A	L	R	A	C	E
H	Y	T	J	R	G	O	X	I	X	A	M	A	A	O
Y	P	S	S	P	T	F	Y	E	O	U	A	Y	R	U
L	O	C	S	Y	Z	I	L	N	N	X	R	N	E	T
O	L	L	O	M	S	Z	I	T	O	E	G	E	N	G
G	Y	A	R	G	E	L	M	I	M	S	O	G	E	R
E	N	D	C	I	I	U	A	F	Y	A	D	O	G	O
N	O	I	T	A	C	I	F	I	S	S	A	L	C	U
E	M	S	U	N	E	G	M	C	M	U	L	Y	H	P
T	I	T	O	M	P	D	D	N	A	O	C	H	S	Y
I	A	I	I	Z	S	C	L	A	D	E	N	P	L	M
C	L	C	F	B	D	X	A	M	G	D	N	I	F	E
V	S	S	C	I	T	A	M	E	T	S	Y	S	B	R
P	U	O	R	G	N	I	M	O	D	G	N	I	K	I

Asexual
Binomial system
Clade
Cladistics
Cladogram
Class
Classification
Ecological race
Ecotypes
Family
Genera
Genus
Ingroup
Kingdom
Outcross
Outgroup
Phylogenetic
Phylogeny
Phylum
Polynomials
Scientific name
Species
Systematics
Taxon
Taxonomy

Crossword

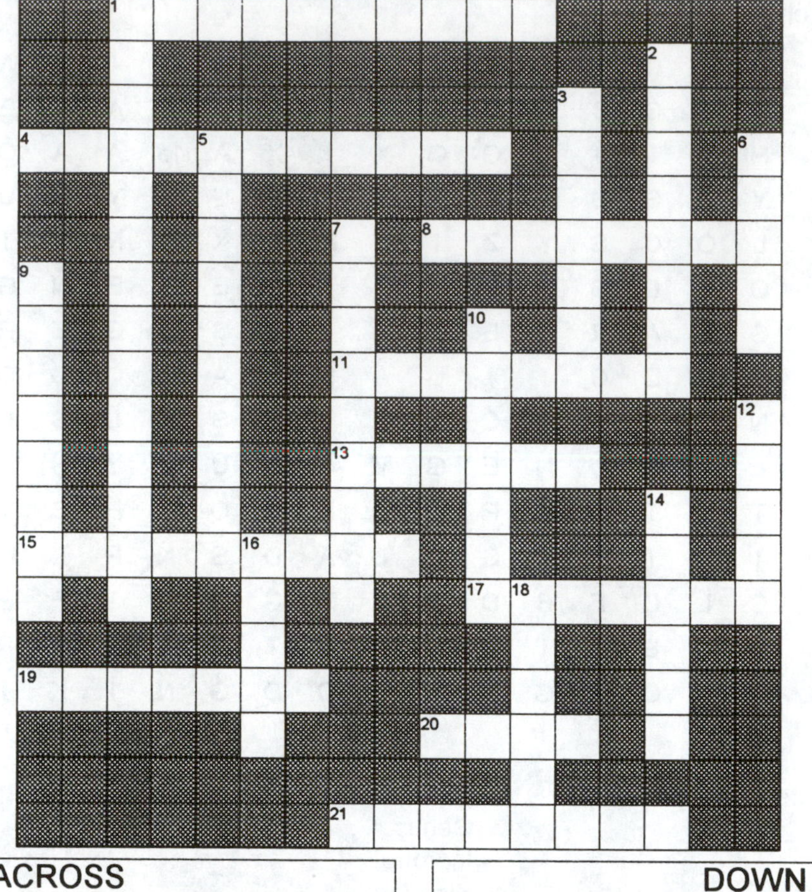

ACROSS

1. A way to construct an evolutionary tree
4. Reconstruction and study of evolutionary trees
8. A common name for the prokaryotes
11. I'll bet his closet was well organized
13. More than one genus
15. ______ mechanisms play a role in speciation
17. The smallest taxonomic category
19. The largest taxonomic category
20. A hybrid of donkeys and horses
21. Two names

DOWN

1. Multilevel grouping of organisms
2. When organisms interbreed with others
3. A population adapted to a local environment
5. The "true" bacteria
6. A group of organisms related by descent
7. Evolutionary relationships
9. Kingdom of multicellular, eukaryotic organisms with no cell walls
10. Bees are these
12. Genus name of the dog
14. Arthropods have an exoskeleton made of this
16. A group of organisms at a particular level
18. A group of similar classes

11 The First Single-Celled Creatures

Chapter Concepts

Bacteria are the most ancient form of life on earth. Bacteria are small, simply organized, single cells that lack an organized nucleus. Bacteria reproduce by binary fission. Some bacteria transfer plasmids from one cell to another in a process called conjugation.

There are two kingdoms of bacteria, archaebacteria and eubacteria. The two kingdoms differ in fundamental aspects of their structure and are found living in very different places. Bacteria are responsible for much of the world's photosynthesis and all of its nitrogen fixation.

Viruses are not organisms; they cannot reproduce outside of cells. Every virus has the same basic structure: a core of nucleic acid encased within a sheath of protein. Animal viruses enter cells by endocytosis, while bacterial viruses inject their nucleic acid into host bacterial cells.

Multiple Choice

1. Oxygen began to accumulate in the early earth atmosphere when
   a. methanogens began to colonize the ocean.
   b. cyanobacteria evolved.
   c. plants evolved.
   d. fungi evolved.
   e. viruses evolved.

2. A major difference between prokaryotic and eukaryotic cells is that
   a. most prokaryotes lack a cell wall, while most eukaryotes have one.
   b. all prokaryotes lack a nucleus, while most eukaryotes have one.
   c. all prokaryotes lack a nucleus, while all eukaryotes have one.
   d. only the prokaryotes can reproduce asexually.
   e. both a and d.

3. Gram-negative bacteria have what advantage over gram-positive bacteria?
   a. Gram-negative bacteria are better able to survive environmental changes than are gram-positive bacteria.
   b. Gram-negative bacteria are able to survive longer without water than can gram-positive bacteria.
   c. Gram-negative bacteria can metabolize a wider variety of nutritional materials than can gram-positive bacteria.
   d. Gram-negative bacteria are able to resist antibiotics that attack the cell wall while gram-positive bacteria are susceptible to these drugs.
   e. Gram-negative bacteria have longer life cycles than do gram-positive bacteria.

4. Endospores function to
   a. help the organism survive periods of drought.
   b. transmit genetic information from one bacterium to another.
   c. help the organism survive exposure to high temperatures.
   d. all of the above.
   e. both a and c.

5. At the end of conjugation, both of the cells involved
   a. form endospores.
   b. perish.
   c. develop organelles.
   d. have a copy of the plasmid DNA.
   e. are able to degrade most antibiotics.

6. HIV infects which of the following cells?
   a. suppresser T-cells
   b. T4 lymphocytes
   c. neutrophils
   d. basophils
   e. platelets

7. Glycoproteins are
   a. proteins with attached sugars that are a component of animal virus spikes.
   b. factors that initiate viral DNA replication.
   c. the product of viral infection by the host cell.
   d. released by infected animal cells to help neighboring cells resist the virus.
   e. none of the above.

8. The end result of the viral lytic cycle is
   a. lysis of the infected cell.
   b. immunity of the host cell to further viral infection.
   c. the release of new viruses.
   d. both a and b.
   e. both a and c.

9. A common cause of the common cold is the
   a. influenza virus.
   b. polio virus.
   c. rhino virus.
   d. HIV.
   e. ebola virus.

Matching

1. __________ Capsule
2. __________ Pili
3. __________ Gram-positive
4. __________ Flagella
5. __________ Endospore
6. __________ *Mycobacterium*
7. __________ Methanogens
8. __________ Heterotroph
9. __________ Chemoautotroph
10. __________ *Anabaena*

a. Require preformed organic material for energy
b. Genus name of a cyanobacterium
c. The cause of tuberculosis
d. They use hydrogen to reduce carbon dioxide
e. They obtain energy from inorganic chemicals
f. They play a role in attachment of bacteria to surfaces
g. A gelatinous covering on some bacteria
h. A membrane made of lipopolysaccharide
i. Thick walls around some cytoplasm
j. Long strands that provide bacterial motility

Completion

1. Motile bacteria possess one or more __________ .
2. Bacteria that are able to thrive in the sulfur springs of Yellowstone National Park are referred to as the __________ .
3. Nonfilterable, infectious agents are called the __________ .
4. HIV binds to __________ on human white blood cells during the infection process.

Cool Places on the 'Net

A Glossary of Microbiology is available at:

http://www.bio.umass.edu/mdid/glossary/p.htm

The American Society for Microbiology provides information on careers in microbiology at:

http://outcast.gene.com/ae/RC/ASM/microbiology_careers.html

Label the Art

A. Label the parts of the bacterial cell.

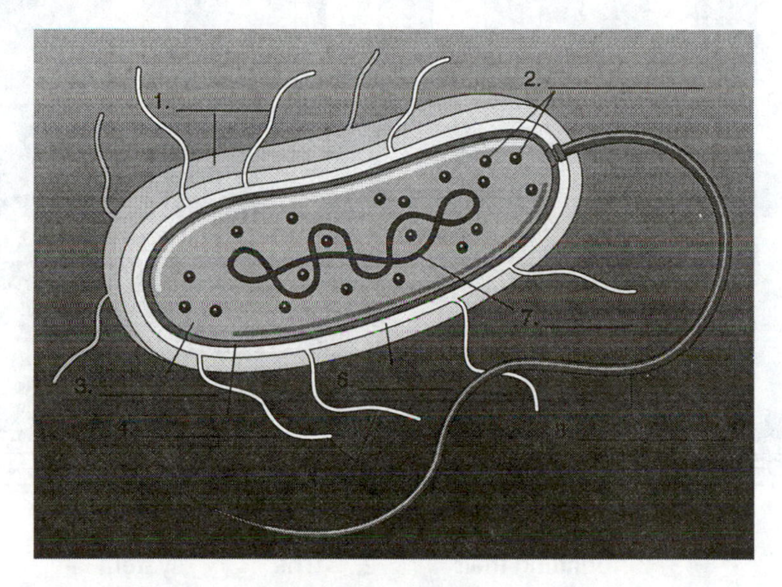

B. Label the parts of the bacteriophage.

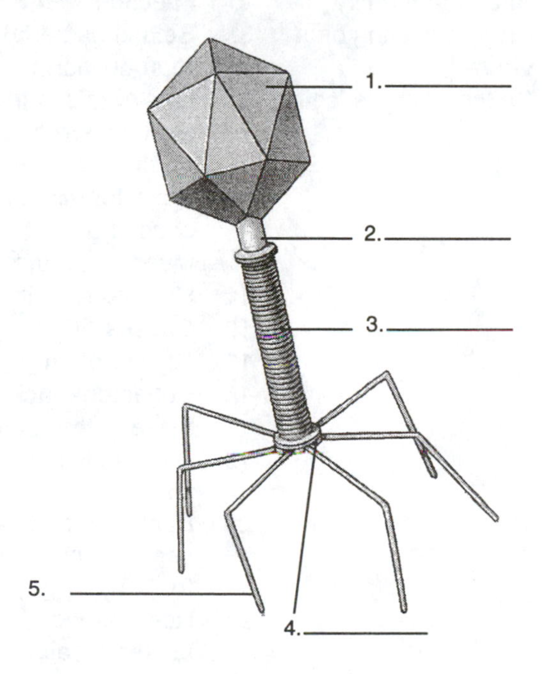

Crossword

ACROSS

1. Gelatinous layer surrounding some bacteria
5. He developed a method to stain bacteria
6. Some bacteria produce resistant ________
7. Not really alive but certainly a problem
8. Round bacterial cell
9. The ancient bacteria are called the _____bacteria
14. Virus that infects E. coli
15. If nitrogen is broken get a bacterium to do this to it
17. Enlarged cells in cyanobacteria
19. Most bacteria have a cell _____ surrounding their plasma membrane
22. Bacterial motility
23. Protein coating on a virus
24. Coiled bacterial cells

DOWN

1. Photosynthetic bacteria
2. The _____ system helps us to resist infection
3. Sexual bacterial reproduction
4. Lots of this in the atmosphere of early earth
10. Short for disease caused by Mycobacterium
11. Hereditary unit
12. Causes AIDS
13. Glycoprotein ______
16. Eubacteria lack these in their genes
18. Surrounds some viruses
20. Bacteria have been common on earth for 3.5 _____ years
21. Rod-shaped bacterial cells
22. Influenza is known as this

12 Advent of the Eukaryotes

Chapter Concepts

The theory of endosymbiosis, accepted by almost all biologists, proposes that mitochondria and chloroplasts were once aerobic eubacteria that were engulfed by ancestral eukaryotes. There is some suggestion that centrioles may also have an endosymbiotic origin.

Of the six kingdoms, Protista is by far the most diverse. Protista is a catchall kingdom containing all eukaryotes that are not fungi, plants, or animals. Multicellularity has evolved among the protists many times.

Reproduction without sex is the rule among protists, which typically resort to sexual reproduction only in times of stress. Sex appears to have first evolved as a mechanism to repair damage to DNA.

The 14 major phyla of protists can be sorted into seven general groups, based largely on their mode of locomotion. Algae are photosynthetic protists, many of which are multicellular. Molds are heterotrophs with restricted mobility.

Multiple Choice

1. The first fossils of protozoans are approximately _________ years old.
   a. 1.5 million
   b. 1.5 billion
   c. 3.5 million
   d. 3.5 billion
   e. 4.0 million

2. Mitochondria probably evolved from
   a. ancestors of the phylum Chlorophyta.
   b. aerobic forams.
   c. anaerobic, fermentative bacteria.
   d. aerobic, symbiotic eubacteria.
   e. archaebacteria.

3. Endosymbiosis
   a. was only able to occur in the earth's early, oxygen-free atmosphere.
   b. probably led to all modern lines of eukaryotic organisms.
   c. has been disproved in several significant experiments.
   d. resulted in the division of the bacteria into two groups: the Eubacteria and the Archaebacteria.
   e. was unable to tolerate the pressures of natural selection.

4. The photosynthetic pigments found in red algae most closely resemble
   a. *Prochloron.*
   b. the green algae.
   c. modern plants.
   d. cyanobacteria.
   e. the euglenoids.

5. Which of the following is a type of asexual reproduction?
   a. parthenogenesis
   b. conjugation
   c. mitosis
   d. fertilization
   e. both c and d

6. You are trying to develop a new automotive paint and you want it to be eye-catching. To add sparkle to the paint you might add
   a. pigments extracted from brown algae.
   b. powdered dinoflagellates.
   c. diatomaceous earth.
   d. coral.
   e. none of the above.

7. While cataloging the life-forms in a sample of ocean water you observe some interesting microscopic organisms. They are present in a variety of shapes; some are round while others have an elongated shape. They all have highly refractive shells, and after running some tests you determine the shells are made up of calcium carbonate. These organisms are probably members of the phylum
   a. Zoomastigophora.
   b. Dinoflagellata.
   c. Bacillariophyta.
   d. Rhizopoda.
   e. Sarcodina.

8. In the famous horror movie *The Blob,* the monster had many features in common with the phylum
   a. Acrasiomycota.
   b. Ciliophora.
   c. Oomycota.
   d. Myxomycota.
   e. Sporozoa.

9. The sporozoans are
   a. motile by cilia and mostly nonparasitic.
   b. motile by rhizopodia and a few species are parasitic.
   c. nonmotile in the haploid state and parasitic.
   d. motile by flagella and often form colonies.
   e. nonmotile and parasitic.

10. Which of the following is being explored as a way to eliminate malaria?
    a. the development of a vaccine
    b. the development of methods to eliminate the *Anopheles* mosquito
    c. the development of effective drugs
    d. all of the above
    e. both a and b

Matching

1. __________ Rhodophyta
2. __________ Rhizopoda
3. __________ Bacillariophyta
4. __________ Phaeophyta
5. __________ Chlorophyta
6. __________ Foraminifera
7. __________ Dinoflagellata
8. __________ Zoomastigina
9. __________ Euglenophyta
10. __________ Ciliophora

a. *Ulva* and *Volvox* are members of this group
b. This group may be an ancestor to animals, and they all have at least one flagellum
c. Sexual reproduction by conjugation occurs in this group
d. They may have a coat of cellulose and silica
e. This group uses pseudopods; *Amoeba* is an example
f. They have flagella, chloroplasts, and paramylon granules
g. Also called the diatoms
h. Many have brightly colored calcium carbonate shells
i. This group is thought to be the most ancient eukaryotes
j. They are also called the brown algae; kelp is an example

Completion

1. Plant life cycles include haploid and diploid phases and are called __________ .
2. The __________ use podia to swim and capture food.
3. The __________ mosquito transmits the __________ of *Plasmodium.*

Cool Places on the 'Net

Information about the phylogeny of the kingdom Protista is available at:

http://www-museum.unl.edu/asp/lifetree.html

Label the Art

A. Label the parts of *Euglena.*

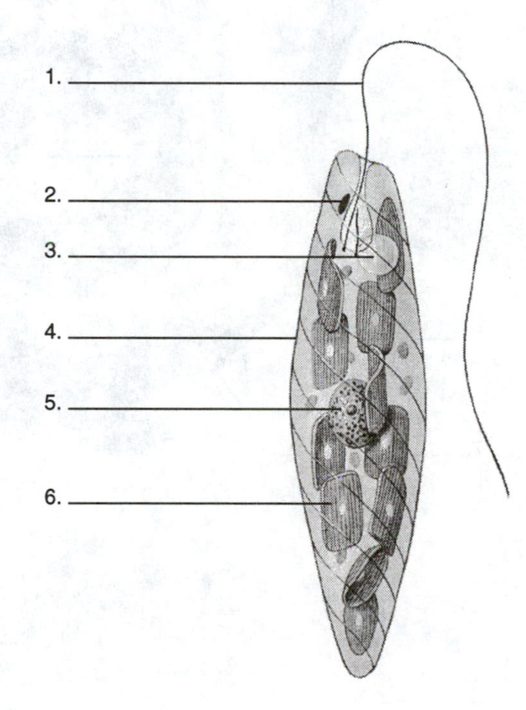

B. Complete the life cycle of the protozoan parasite *Plasmodium.*

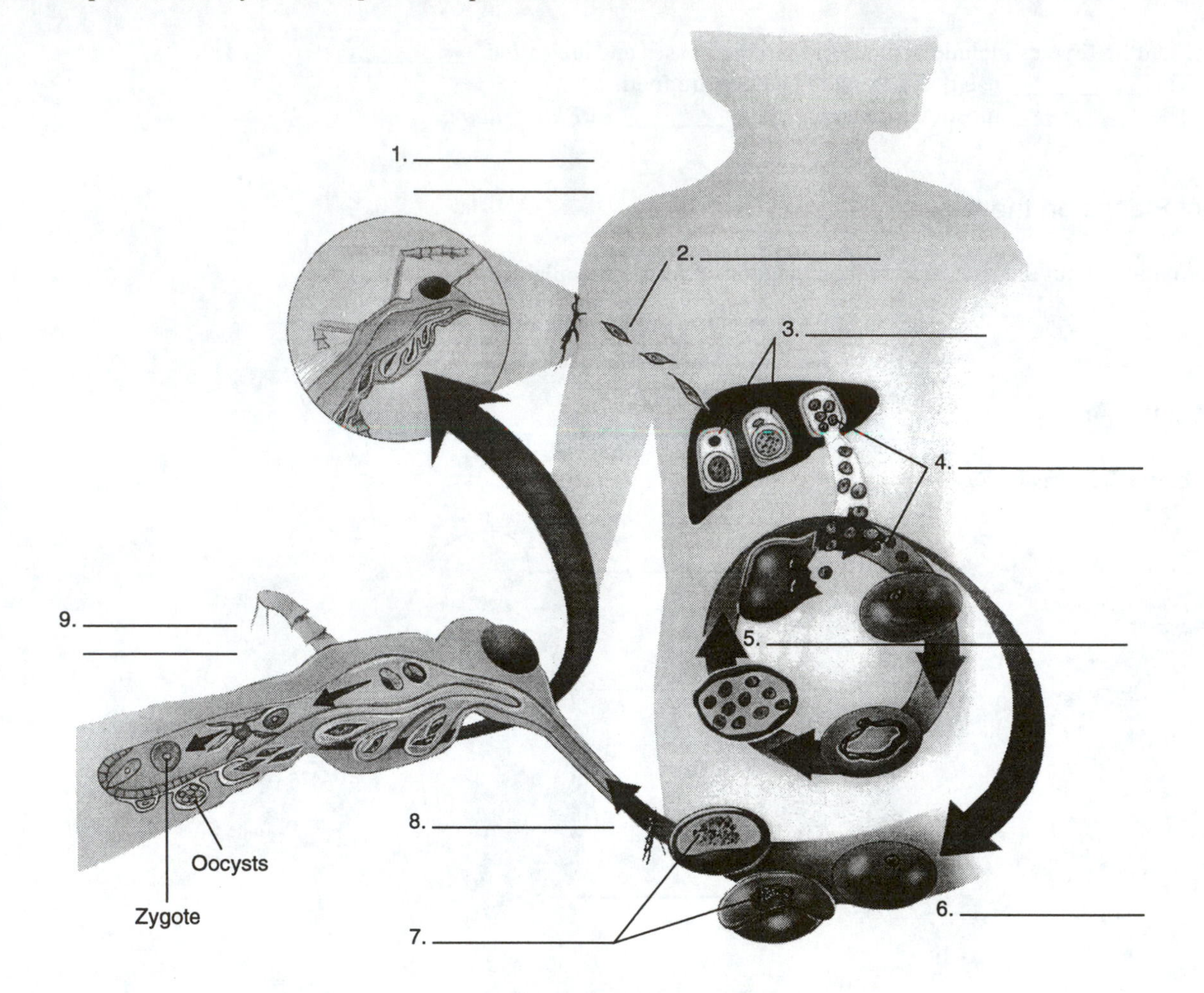

Crossword

ACROSS

2. On the fence between plants and animals
6. A disease caused by a sporozoan
8. Early eukaryotic cells may have been a host in an endosymbiotic relationship
9. Aerobic respiration
12. This covers many protozoans
14. Another one that probably arose from an endosymbiotic relationship
16. Bioluminescent ocean dweller
18. They move by cytoplasmic streaming
20. Move with "false feet" and reproduce by fission
21. Motile by means of pseudopods

DOWN

1. A sporozoan
2. They are the "real" thing
3. A light-sensitive organ
4. A protozoan that causes disease in humans
5. Plants have them but animals do not
7. Spreads a sporozoan that causes disease in humans
10. Dinoflagellates have a stiff coat of this
11. Photosynthetic and unicellular
13. Photosynthetic boxes with lids
15. The contractile ______ plays a role in water regulation
17. This comes from red algae
19. Phycobilins are this color

13 Evolution of Multicellular Life

Chapter Concepts

Multicellular organisms are composed of many cells that integrate their activities. Colonial organisms are collections of cells that are permanently associated but in which little integration of cell activities occurs. Aggregations are transient collections of cells.

Simple multicellular organisms lack complex cell specialization.

Complex multicellular organisms are characterized by cell specialization and intercell coordination.

Fungi are complex multicellular heterotrophs with filamentous bodies. Long strings of fungal cells, called hyphae, form a mass called a mycelium. The cells of fungi are interconnected, sharing cytoplasm and nuclei.

Lichens are symbiotic associations between a fungus and a photosynthetic partner. Mycorrhizae are symbiotic associations between a fungus and the roots of a plant.

Multiple Choice

1. The evolution of multicellularity
   a. was necessary to overcome surface-to-volume problems that larger cells would have encountered.
   b. was an advantage in avoiding prokaryotic predators.
   c. allowed organisms to more efficiently utilize oxygen.
   d. occurred prior to the evolution of plants.
   e. both a and d.

2. Which of the following represents an aggregation?
   a. kelp
   b. hydra
   c. *Volvox*
   d. Cellular slime mold
   e. lichen

3. The green algae were once classified as plants but now are in the kingdom Protista. Why is this?
   a. The green algae have several animal-like characteristics.
   b. The green algae are mostly aquatic and are simpler than the plants.
   c. Although the green algae have photosynthetic pigments, they lack chloroplasts.
   d. The green algae tend to form colonies.
   e. The green algae represent a separate evolutionary line.

4. While walking through the woods in New England, you notice a crusty material on several of the rocks. You observe a slice of the material with a light microscope and notice long filamentous strands of cells, many of which surround round nucleated cells. What might this be?
   a. a mold
   b. a colony of bacteria
   c. a lichen
   d. an amoeba
   e. a mushroom

5. The __________ were probably the ancestors of modern plants.
   a. protozoans such as amoeba
   b. green algae
   c. red algae
   d. cellular slime molds
   e. ciliates

6. The fungi probably evolved from
   a. plantlike protists.
   b. the green algae.
   c. the plants.
   d. protozoans.
   e. something unknown at this time.

7. A person who studies fungi is called a(n)
   a. ornithologist.
   b. symbiosis scientist.
   c. mycologist.
   d. botanist.
   e. none of the above.

8. A characteristic that is unique to the fungi is
   a. nuclear mitosis.
   b. binary fission.
   c. the presence of chitin.
   d. absorption of nutrients.
   e. the presence of cellulose.

9. A heterokaryon is
   a. any heterozygous organism.
   b. an asexual spore produced by members of the phylum Basidiomycota.
   c. a fungal cell containing genetically different nuclei.
   d. a diploid, nonseed plant.
   e. not viable under most circumstances.

10. Fungi obtain their nutrients by
   a. photosynthesis.
   b. the ingestion of bacteria and protozoans.
   c. internal digestion.
   d. external digestion.
   e. both a and b.

11. Most of the fungi that are human pathogens are members of the phylum
   a. Zygomycota.
   b. Ascomycota.
   c. Basidiomycota.
   d. Fungi imperfecti.
   e. none of the above.

12. Which of the following fungi do not produce a heterokaryon upon fusion of hyphae of different mating strains?
   a. zygomycetes
   b. ascomycetes
   c. basidiomycetes
   d. Fungi imperfecti
   e. yeasts

13. Some members of this phylum are important agricultural pests while others are delicious.
   a. Zygomycota
   b. Ascomycota
   c. Basidiomycota
   d. Fungi imperfecti
   e. yeasts

14. The antibiotic penicillin is produced by a species in the phylum
   a. Zygomycota.
   b. Ascomycota.
   c. Basidiomycota.
   d. Fungi imperfecti.
   e. yeast.

15. A fungus that has septa and that usually reproduces asexually is a member of phylum
   a. Zygomycota.
   b. Ascomycota.
   c. Basidiomycota.
   d. Fungi imperfecti.
   e. yeast.

16. An ascus is
    a. an asexual spore produced by members of Basidiomycota.
    b. a sexual spore produced by yeasts when environmental conditions are adverse.
    c. a structure in which the zygote forms in members of phylum Ascomycota.
    d. the result of a nutritional deficiency.
    e. both a and d.

17. The holes in septa function to
    a. allow the fungus to absorb nutrients.
    b. allow for cell-to-cell communication.
    c. produce asexual spores.
    d. allow for sexual reproduction.
    e. none of the above.

18. In lichens the fungal partner provides
    a. protection to the photosynthetic partner from the environment.
    b. carbohydrates to be used for nutrition.
    c. minerals and other nutrients.
    d. all of the above.
    e. both a and c.

19. Approximately __________ species of __________ are involved in all endomycorrhizae worldwide.
    a. 30, zygomycetes
    b. 300, ascomycetes
    c. 3,000, basidiomycetes
    d. 30,000 Fungi imperfecti
    e. none of the above

Completion

1. The cells that make up *Volvox* represent a(n) __________ organism.
2. The process by which a single cell becomes a multicellular organism is called __________ .
3. Long filaments of fungal cells are called __________ . A mass of these filaments is called a(n) __________ .
4. The __________ have the smallest number of species of the fungal phyla.
5. Fusion of hyphae of different mating strains of members of phylum Basidiomycota produces a(n) __________ .
6. A basidiocarp can also be referred to as a(n) __________ .

Cool Places on the 'Net

Fungi Perfecti offers books and supplies to grow a variety of gourmet mushrooms at:

http://www.halcyon.com/mycomed/fppage.html

An "Introduction to the Fungi" and a multimedia glossary is available at:

http://ucmpl.berkeley.edu/fungi/fungi.html

Label the Art

Complete the life cycle of a zygomycete.

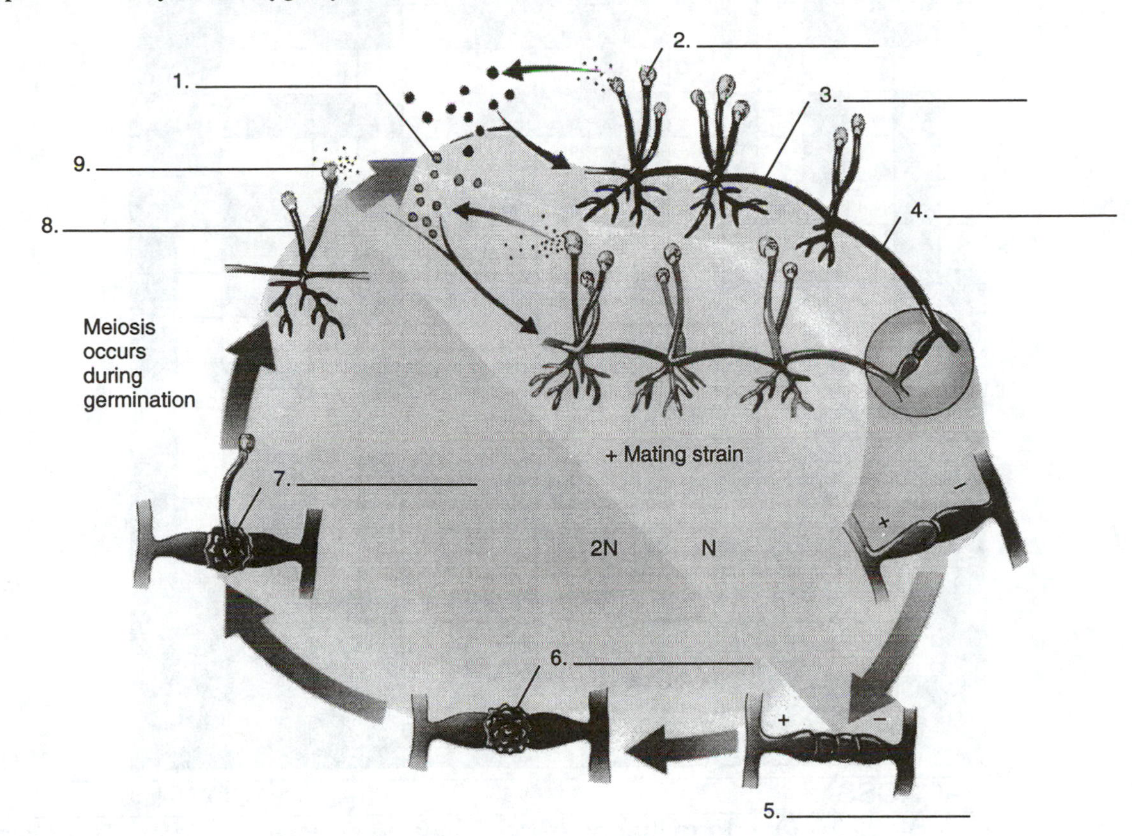

Crossword

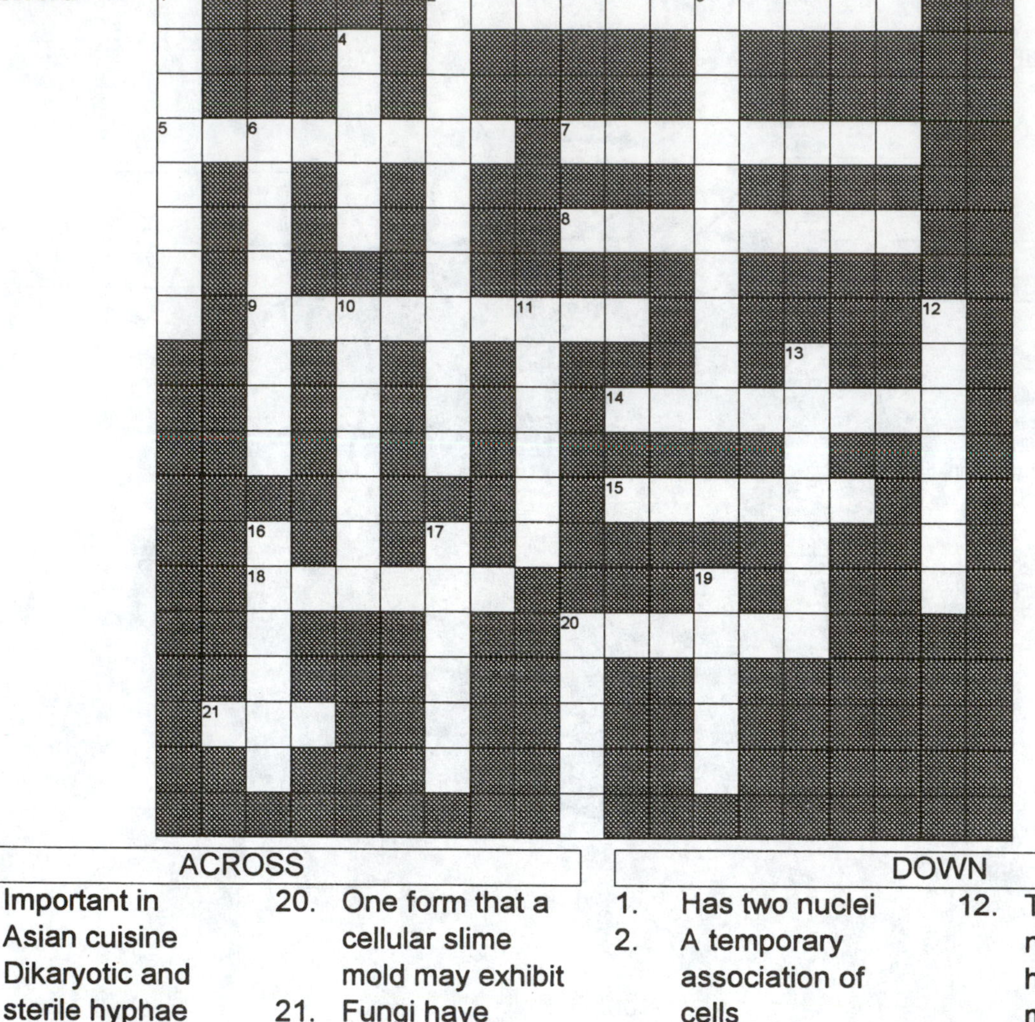

ACROSS

2. Important in Asian cuisine
5. Dikaryotic and sterile hyphae
7. A woven mat of fungal strands
8. Common name for one member of phylum Basidiomycota
9. Pleurotus snacks on these
14. This fungus is a member of phylum Zygomycota
15. This is a component of fungi cell walls
18. This is a tasty mushroom
20. One form that a cellular slime mold may exhibit
21. Fungi have ___-motile sperm

DOWN

1. Has two nuclei
2. A temporary association of cells
3. The Fungi ________ lack sexual spores
4. This forms when hyphae from two fungi fuse
6. A permanent collection of associated cells
10. The fungi undergo nuclear ________
11. A brown kelp may grow to ________ of meters in length
12. The fusion of nuclei occurs here in mushrooms
13. A type of asexual fungal spore
16. It is almost multicellular
17. They help bread rise
19. Walls that divide fungal cells
20. The photosynthetic partner in a symbiotic relationship

14 Rise of the Flowering Plants

Chapter Concepts

Plants faced three key challenges in adapting to life on land: absorbing minerals, conserving water, and transferring gametes during sexual reproduction. Early plants developed a symbiotic partnership with fungi to aid in absorbing minerals. Plants conserve water with a watertight outer covering called a cuticle. Plants transfer protected gametes, as spores or pollen. Plants exhibit alternation of gametophyte (haploid) and sporophyte (diploid) generations.

Vascular plants possess specialized water-conducting tissues. The life cycle of vascular plants is dominated by the sporophyte generation. Early vascular plants grew from their tips (primary growth), while later ones also grow in diameter (secondary growth).

A seed is a plant embryo with a durable, watertight cover. In the gymnosperms, the ovules are not completely enclosed within sporophyte tissue when pollinated. Seeds often contain nourishment for the germinating seedling.

Angiosperm ovules are completely enclosed by the sporophyte carpel when pollinated. Angiosperms, or "flowering plants," are the most successful of all plants. Angiosperms are characterized by flowers and fruit, both of which aid dispersal of gametes.

Multiple Choice

1. Which of the following probably made the move to land by plants possible?
   a. the advent of multicellularity
   b. the accumulation of methane in the atmosphere
   c. the development of an ozone layer
   d. the development of the seed
   e. evolution of a flagellated sperm

2. Mycorrhizae
   a. enable a plant to photosynthesize more efficiently.
   b. are lichens.
   c. enable a plant to absorb many essential minerals.
   d. are respiratory structures found in some plants.
   e. both a and d.

3. Water conservation in plants is assisted by
   a. the cuticle.
   b. transpiration.
   c. stomata.
   d. cilia.
   e. both a and c.

4. The haploid stage in a plant life cycle is called the
   a. zygote.
   b. embryo.
   c. gametophyte.
   d. sporophyte.
   e. conspicuous stage.

5. A redwood tree represents which stage in the life cycle of that plant?
   a. zygote
   b. embryo
   c. gametophyte
   d. sporophyte
   e. none of the above

6. A difference between the gametophytes of nonvascular and vascular plants is that
   a. the nonvascular gametophyte is haploid, while the vascular gametophyte is diploid.
   b. the nonvascular gametophyte is bisexual, while the vascular plants have separate male and female gametophytes.
   c. the nonvascular gametophyte can only grow in water, while vascular gametophytes can grow on land.
   d. in nonvascular plants the gametophyte is the inconspicuous stage, while it is the conspicuous stage in vascular plants.
   e. there is no major difference between them.

7. One characteristic that modern mosses share in common with early plants is the
   a. need for water during fertilization.
   b. pollen grain.
   c. lack of a diploid stage.
   d. need for insects or wind for pollination.
   e. protected seed.

8. Why was vascular tissue such an important evolutionary advance?
   a. It made it easier for fertilization to occur on dry land.
   b. It allowed plants to have protected gametophytes.
   c. It was necessary for the formation of a seed coat.
   d. It allowed plants to grow larger and still conduct water to all of the tissue.
   e. Both a and d.

9. If the sporophyte was removed from the gametophyte in a Bryophyte, what would happen?
   a. The sporophyte would die because it is dependent on the gametophyte for nutrition.
   b. The gametophyte would die because it is dependent on the sporophyte for nutrition.
   c. The sporophyte would be able to survive on its own and it would produce another gametophyte.
   d. The sporophyte would form a protective capsule to help it survive until adequate nutrients were available.
   e. The sporophyte would continue to undergo meiosis.

10. Which of the following lack vascular tissue?
    a. moss
    b. the gymnosperms
    c. the hornworts
    d. angiosperms
    e. the monocots

11. The cyads were the predominant land plants during the
    a. Cooksonian.
    b. Jurassic.
    c. Precambrian.
    d. Cambrian.
    e. none of the above.

12. The endosperm functions to
    a. protect the seed from desiccation.
    b. keep the seed dormant until water is available.
    c. provide nutrients to the young plant embryo.
    d. produce pollen grains.
    e. produce vascular tissue in the plant after the seed germinates.

Matching

1. __________ Frond
2. __________ Meristem
3. __________ Phloem
4. __________ Xylem
5. __________ Wood
6. __________ Seed
7. __________ Spore

a. Transports carbohydrates made in the leaves to other parts of the plant
b. A plant embryo
c. A fern leaf
d. Fern sporophytes release these
e. Actively growing cells at the tip of the plant
f. Transports water in vascular plants
g. This is the result of secondary plant growth

Completion

1. Organisms that are able to produce carbohydrates by photosynthesis are called __________ .
2. In vascular plants the male gametophyte is also called a(n) __________ .
3. Redwoods are members of phylum __________ .

Cool Places on the 'Net

Check out the World Wide Web Virtual Library on plant biology at:

http://golgi.harvard.edu/biopages/botany.html

Label the Art

A. Complete the generalized plant life cycle.

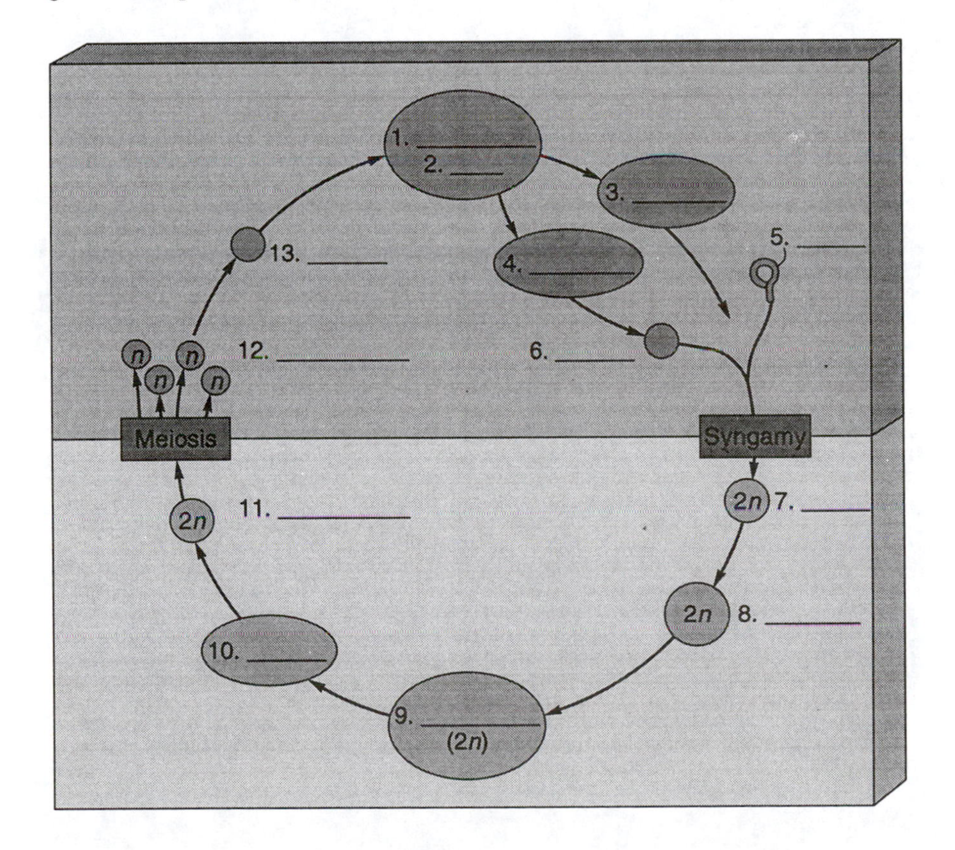

B. Label the parts of a flower.

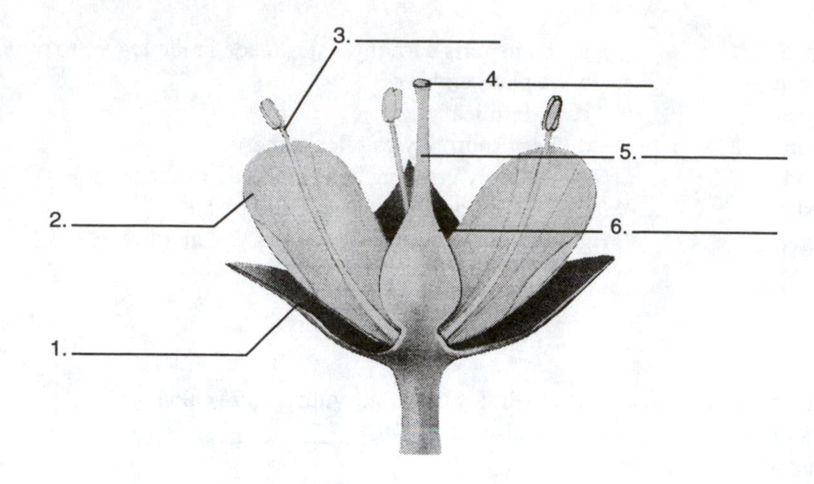

Crossword

ACROSS	DOWN

ACROSS

1. A revolution in plant reproduction
3. _______ cells flank an opening in the epidermis of plants
7. When oxygen began to accumulate in the earth's early atmosphere this began to form
8. An actively dividing zone of cells in a plant
9. The ______ stage of a plant is diploid
11. Port into and out of a plant
12. They are seedless vascular plants
14. They lack vascular tissue and have a funny name
17. Fern leaves
18. One of six inorganic minerals that plants require
20. They are required to build strong bodies
21. The mosses require this for fertilization
23. A symbiotic relationship between a plant and a fungus
24. The reproductive organs of angiosperms

DOWN

1. The fern life cycle includes these
2. A highly nutritious tissue
3. Naked seed plants are this
4. Apple is this kind of fleshy fruit
5. A low-growing plant with vascular tissue
6. This kind of tissue conducts carbohydrates in plants
10. _______ tissue conducts water and mineral in plants
13. A covering that helps plants retain water
15. Wood is this type of growth
16. The male gametophyte
19. Encloses the ovule
20. Corn is an example
22. This is how many phyla of modern plants are vascular

15 The Living Plant

Chapter Concepts

A vascular plant is organized along a vertical axis. The part aboveground is called the shoot and encompasses the stem, its branches, and the leaves. The part belowground is called the root. Plants grow from actively dividing zones called meristems. Growth from the tip of the shoot or root is called primary growth. Growth in girth is called secondary growth. It takes place in a meristem called the vascular cambium.

Water is held up in the conducting vessels of plants by adhesion to the vessel walls and cohesion of the water molecules to one another. Water is pulled up the plant by transpiration, which is evaporation from the surface of leaves. Carbohydrates are translocated throughout the plant by mass flow. Auxin is a hormone that regulates cell growth in plants.

The male and female reproductive structures occupy the third and fourth whorl of a flower. If the male and female structures mature at the same time, the flower may self-pollinate. If the male and female structures mature at different times, or on different flowers, the pollen grains from the male structures are carried to the female structures of different flowers by insects or by the wind.

Multiple Choice

1. Which of the following cell types has no living cytoplasm at maturity?
   a. tracheids
   b. sieve-tube members
   c. companion cells
   d. collenchyma
   e. meristems

2. Which of the following cell types lacks nuclei at maturity?
   a. tracheids
   b. sieve-tube members
   c. companion cells
   d. collenchyma
   e. meristems

3. Which of the following cell types retains the ability to differentiate into another cell type if the need should arise in a plant?
   a. tracheids
   b. sieve-tube members
   c. companion cells
   d. collenchyma
   e. meristems

4. Pits allow water to move through
   a. vessel elements.
   b. tracheids.
   c. companion cells.
   d. sieve-tube members.
   e. both a and d.

5. Leaf growth is due primarily to __________ meristems.
   a. apical
   b. marginal
   c. lateral
   d. vertical
   e. horizontal

6. Large intercellular spaces in which gas exchanges occur between
   a. meristems.
   b. stipules.
   c. spongy parenchyma.
   d. ground tissue.
   e. cortex.

7. Periderm is
   a. cork cells.
   b. a dense layer of parenchyma.
   c. the cork cambium.
   d. all of the above.
   e. none of the above.

8. While collecting firewood to heat their mountain cabin, a family finds a dead juniper tree that has fallen over. After cutting the trunk into lengths, they notice the annual rings and count them. There are 34 rings. This means that the tree is
   a. 34 years old.
   b. approximately 34 years old.
   c. 340 years old.
   d. 17 years old.
   e. The growth rings do not really reflect age so it is not possible to determine the age of the tree.

9. A plant with damaged root hairs would
   a. be unable to absorb water.
   b. be unable to absorb minerals.
   c. rely on the root for water absorption.
   d. both a and b.
   e. both b and c.

10. Tensile strength can be increased by
    a. increasing the diameter of the tube of water.
    b. decreasing the diameter of the tube of water.
    c. adding an emulsifier to the water.
    d. increasing the temperature of the water.
    e. decreasing the temperature of the water.

11. While preparing for a dinner party at your house, you cut up some celery to eat with dip. You get busy and forget to put the celery in the refrigerator for awhile and it becomes slightly soft. You place the celery in the refrigerator in some water and an hour later it is once again crispy. Why?
    a. The turgor pressure has been restored.
    b. The cold temperature made the celery crispy.
    c. Water has adhered to the celery, making it crispy.
    d. The cold temperature has activated enzymes that play a role in rigidity.
    e. None of the above.

Matching—Part 1

1. __________ Fibers
2. __________ Tracheids
3. __________ Trichomes
4. __________ Sclereids
5. __________ Petiole
6. __________ Companion cells
7. __________ Parenchyma
8. __________ Fibers
9. __________ Guard cells
10. __________ Collenchyma

a. The least specialized and the most common of plant cell types
b. A slender stalk that is attached to the leaf
c. They are alive at maturity and form strands under the epidermis of stems
d. A type of parenchyma that is associated with sieve-tube members
e. A type of sclerenchyma that forms long strands
f. Long, slender sclerenchyma cells that form strands
g. They have pits and help make up xylem
h. They are variable in shape but are often branched; may be called stone cells
i. Paired cells that make up the stomata
j. These outgrowths of the epidermis may make a leaf look "fuzzy"

Matching—Part 2

1. __________ Cytokinins
2. __________ Abscisic acid
3. __________ Ethylene
4. __________ Gibberellins
5. __________ Auxins

a. This hormone causes leaves to age and fall off of the plant
b. This hormone would cause the stem to grow towards a source of light
c. This hormone is produced in the apical portion of stems
d. Tomatoes are stimulated to ripen when exposed to this hormone
e. Cell division is stimulated by this hormone

Completion

1. The aboveground portion of a plant is called the __________ , while the portion that is below the ground is the __________ .
2. The flattened part of a leaf is called the __________ .
3. Large quantities of water are lost from a plant through the process of __________ .

Cool Places on the 'Net

The Internet Directory for Botany provides information about plants as food, medicine, and material for textiles. This can be located at:

http://www.helsinki.fi/kmus/botecon.html

Label the Art

Label the parts of a plant.

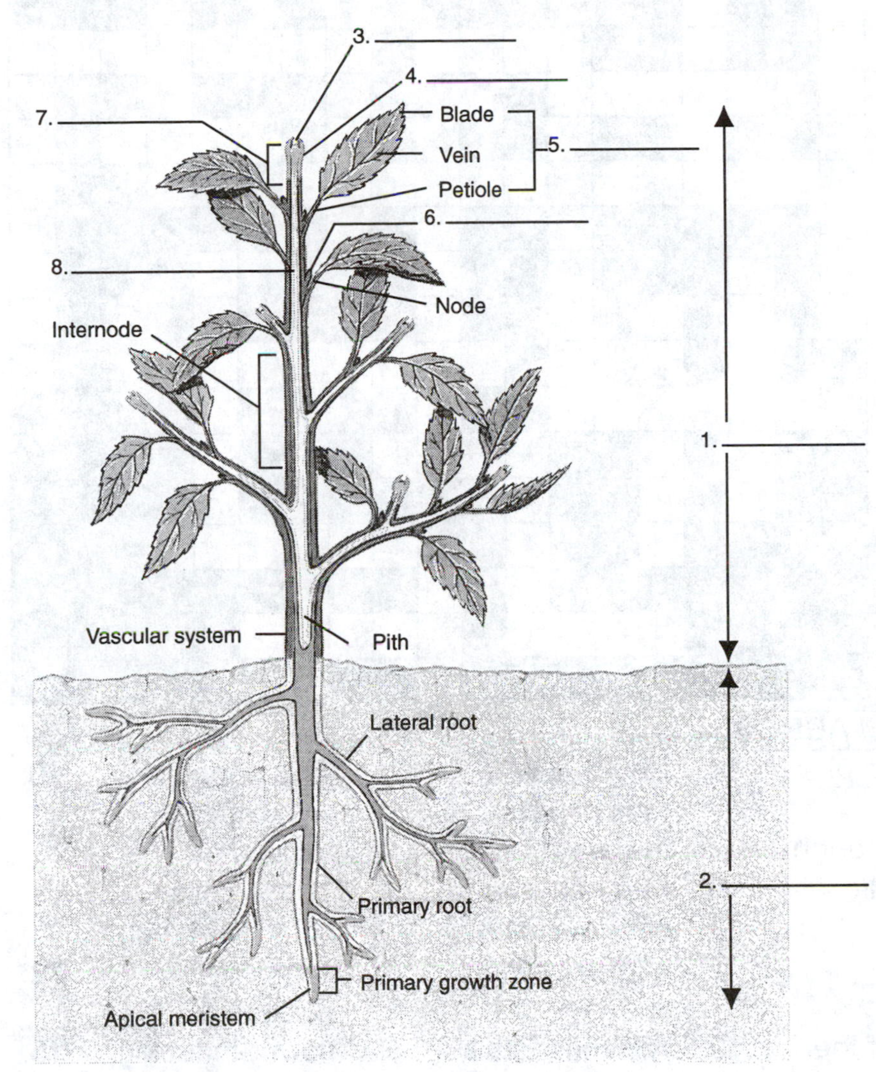

Crossword

ACROSS

1. They have pits that water passes through
5. The "middle leaf"
7. Root _____ absorb the majority of the water
8. The inner portion of a young stem
9. This acid makes leaves fall
11. _____-tube members are found in angiosperms
13. Big-time pollinators
14. Plants lose tons of water this way
18. Parenchyma cells make up a layer called this
19. A slender stalk on a leaf
20. It holds up the plant
22. This tissue covers the plant
24. Absorbs water and minerals
25. It regulates growth in plants

DOWN

2. This happens with water molecules
3. These cells could keep you company
4. Causes a flower-inducing response in plants
6. They have not yet decided what to be when they grow up
8. A cylinder of cells surrounding vascular tissue
10. The vascular kind gives rise to secondary xylem and phloem
12. Produced by ripening fruit
15. A sugary liquid
16. All of the petals
17. ______ parenchyma
20. The part of the plant that is aboveground
21. A plump plant is this
23. One of these is on the end of a growing root

16 Evolution of the Animal Phyla

Chapter Concepts

There are about 36 phyla of animals. Animal cells lack cell walls. All animals are multicellular heterotrophs, and most are mobile.

Metazoans possess tissues and organs, and their bodies have a definite shape and symmetry. Protists called choanoflagellates very closely resemble the choanocytes of sponges and may be the ancestors of all animals. The most primitive metazoans exhibit radial symmetry.

All advanced metazoans are bilaterally symmetrical at some stage of their life cycle. Acoelomate animals have a solid body with no internal cavity. This severely limits their size, since diffusion is the only way they can transport materials.

A pseudocoelom is a body cavity between the endoderm and the mesoderm. A coelom is a body cavity entirely within the mesoderm. It has the advantage that the primary tissues can interact with one another. Segmentation is the building of a body from a series of similar sections. It has the great advantage that different segments can specialize in different ways. A rigid exoskeleton of chitin limits body size because chitin is brittle.

A notochord is a flexible rod to which muscles attach, allowing early chordates to swing their bodies back and forth. The function of the notochord is taken over by the backbone of vertebrates.

Multiple Choice

1. Animal(s)
   a. cells lack cell walls.
   b. sometimes have cell walls, such as in the arthropods.
   c. are always mobile.
   d. are heterotrophic or autotrophic.
   e. are found primarily on land.

2. Choanocytes are
   a. primitive organs in sponges that function in digestion.
   b. flagellated cells that draw water through the body cavity of sponges.
   c. amoeba-like cells that wander over the surface of sponges and distribute nutrients to other cells.
   d. calcium deposits that provide structure to sponges.
   e. both a and b.

3. One property that sponges share in common with many other animals is
   a. a life cycle that involves alternation of generations.
   b. a free-swimming medusa in its life cycle.
   c. cell recognition.
   d. the ability to digest cellulose.
   e. none of the above.

4. Cnidarians project a nematocyst to capture their prey by
   a. ejecting it with a jet of water.
   b. using a springlike apparatus.
   c. building up a high internal osmotic pressure.
   d. employing simple muscle fibers.
   e. coiling and releasing the tendrils on which the nematocysts are found.

5. The cnidarians have which of the following evolutionary advances over the sponges?
   a. bilateral symmetry
   b. cephalization
   c. a life cycle involving a dominant haploid form
   d. extracellular digestion
   e. a nonmotile mature form

6. Which of the following is an example of an organism with the medusae body form?
   a. a sponge
   b. a hydra
   c. a coral
   d. an amoeba
   e. a jellyfish

7. Most cnidarians
   a. exist only as medusa in their life cycle.
   b. exist only as polyps in their life cycle.
   c. encyst when they encounter adverse environmental conditions.
   d. alternate between a polyp and a medusa during their life cycle.
   e. none of the above.

8. Cephalization is
   a. an evolutionary trend in the cnidarians.
   b. the evolution of a head.
   c. the lack of a distinctive head seen in the primitive animal phyla.
   d. the trend towards bilateral symmetry seen in the chordates.
   e. missing in many of the animal phyla.

9. An example of an animal with a pseudocoelomate body is (are)
   a. hydra.
   b. planaria.
   c. rotifers.
   d. Dugesia.
   e. jellyfish.

10. A radula is
    a. a sharp structure that is injected into the prey of a mollusk.
    b. a protective coating made of calcium carbonate on sponges.
    c. a small, internal shell found in cephalopods.
    d. a rasping, tonguelike organ in mollusks.
    e. necessary for mollusks to be motile.

11. The segments of annelids are
    a. actually one large unit internally.
    b. partitioned internally.
    c. specialized for different functions.
    d. not present in all species.
    e. both b and c.

12. Rocky Mountain spotted fever and Lyme disease are caused by a(n)
    a. insect.
    b. protozoan.
    c. fungus.
    d. nematode.
    e. arachnid.

13. How do the crustaceans differ from the insects?
    a. Crustaceans have an exoskeleton made of chitin.
    b. Crustaceans have legs on their abdomen and thorax.
    c. Crustaceans have jointed appendages.
    d. Crustaceans are exclusively found in marine habitats.
    e. Both a and c.

14. Malpighian tubules
    a. assist ecinoderms in motility.
    b. regulate osmotic pressure in insect cells.
    c. play a major role in digestion in insects.
    d. function in excretion in insects.
    e. secrete the exoskeleton in insects.

15. Animals in which the blastopore becomes the mouth are called
    a. stomates.
    b. deuterostomes.
    c. protostomes.
    d. echinostomes.
    e. prestomes.

16. The significance of a notochord in the evolution of chordates is that it
    a. provided an internal attachment point for muscles.
    b. allowed for the development of a more complex nervous system.
    c. eliminated the need for segmentation.
    d. allow the organism to grow larger.
    e. both a and c.

Completion

1. Members of which three animal phyla dominate life on land?

   ____

   ____

   ____

2. The comb jellies are members of the phylum __________ .
3. __________ refers to a solid body construction and is exemplified by the phylum __________ .
4. The development of a body cavity provided what three advantages to animals?

   ____

   ____

   ____

5. The development of specialized tissues in animals involves a process known as __________ .
6. __________ is the animal phylum with the greatest diversity.
7. Insect exoskeletons are made up primarily of __________ and function as __________ .
8. Arthropods lacking jaws are called __________ , while those that have jaws are called __________.
9. The three regions of an insect body are the __________ , __________ , and __________ .
10. Millipedes have __________ pair(s) of legs per body segment and centipedes have __________ pair(s).

Cool Places on the 'Net

To hook up with a variety of web sites that provide medical information, check out the ML Communications home page at:

http://members.gnn.com/mlco

Label the Art

Label the parts of a lobster.

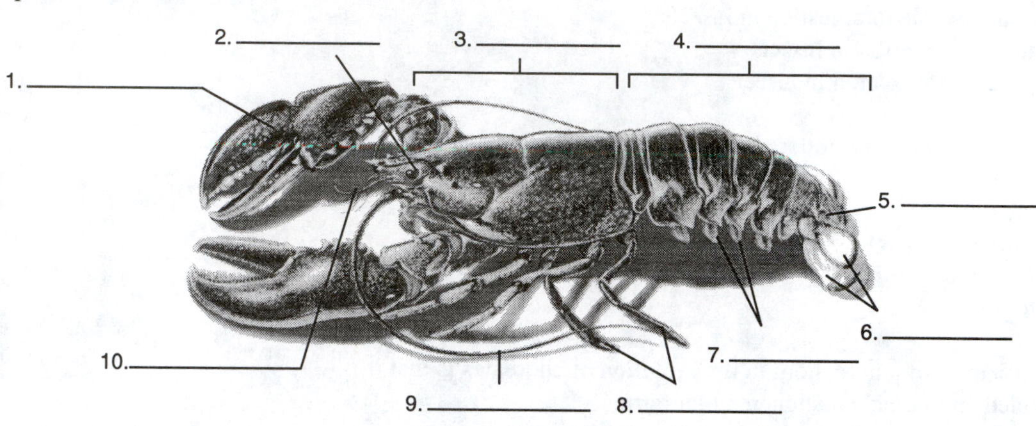

Crossword

ACROSS	DOWN

ACROSS

3. These flagellated cells draw water through a sponge
7. A fluid-filled body cavity
8. The flatworms were the first to have one of these
10. Humans display ___lateral symmetry
11. The top of a fish is the _______ side
12. The backside of an organism
14. The front of an organism
16. They are commonly found in soil
17. A harpoon for cnidarians
21. Most of the flatworms are _______
22. Genus name for the lobster

DOWN

1. These worms are the simplest of the bilaterally symmetrical animals
2. Members of phylum Porifera may be up to 2 _______ long
3. There must be a head up there somewhere
4. This means to have a solid body
5. "Spiny skin"
6. The inner layer of an embryo
9. The nervous system develops from this embryonic layer
13. Earthworms are these
15. Chordates have one of these
18. Jellyfish exhibit this type of body plan
19. They capture oxygen from the water
20. These allow for flight

17 History of Terrestrial Vertebrates

Chapter Concepts

All major groups of living organisms except plants arose in the early Paleozoic era. Arthropods and vertebrates invaded land from the sea soon after plants and fungi. Periodic mass extinctions have occurred; the greatest, at the end of the Paleozoic era, rendered 90% of all species extinct. In the Mesozoic era reptiles were the dominant terrestrial vertebrate, particularly dinosaurs. At the end of the Mesozoic era dinosaurs disappeared, and mammals took their place.

The first vertebrates were jawless fishes. The three key innovations in the evolutionary history of fishes were jaws, fins, and the swim bladder. Amphibians, the first vertebrates to live on land, evolved legs, lungs, and the pulmonary vein. Reptiles were well-adapted to living on land, with dry skin and watertight eggs. Birds evolved feathers, hollow bones, and very efficient lungs. Most present-day mammals are placental mammals.

Marine and aquatic vertebrates extract oxygen from water very efficiently with gills. Reptiles and mammals perfect the lung by increasing its surface area with alveoli. Terrestrial vertebrates evolved a septum within the heart to separate the pulmonary and systemic circulations. Reptiles, birds, and mammals conserve water by using kidneys to remove it from urine.

Multiple Choice

1. Which of the following is the correct order, from largest to smallest?
   a. epochs, eras, periods, ages
   b. ages, eras, epochs, periods
   c. periods, epochs, ages, eras
   d. eras, periods, epochs, ages
   e. ages, epochs, periods, eras

2. During the Cambrian
   a. many new successful forms of animal life evolved, leading to forms that still survive.
   b. a variety of new body forms and lifestyles evolved in animals.
   c. many animal phyla evolved that have no living relatives.
   d. palmlike plants were the dominant form of life.
   e. both b and c.

3. While searching for fossils in the Sonoran Desert, you find some trilobites. You know that these fossils are about __________ million years old.
   a. 500
   b. 250
   c. 100
   d. 55
   e. 25

4. On a fossil-hunting expedition, you find an animal that looks like a cross between a fish and an amphibian. That animal is probably
   a. *Archeopteryx.*
   b. *Ichthyostega.*
   c. a pterosaur.
   d. a plesiosaur.
   e. a trilobite.

5. How did the evolution of a notochord contribute to the invasion of land by animals?
   a. The notochord led to the development of a spinal cord and allowed animals to better respond to external stimuli.
   b. The notochord provides a site for the internal attachment of muscles, allowing for more efficient locomotion.
   c. The notochord led to lung development and a new form of life that could better survive on land.
   d. Both a and c.
   e. The notochord did not contribute to the evolution of land animals.

6. Mass extinctions occurred during the
   a. Cenozoic.
   b. Silurian.
   c. Triassic.
   d. Carboniferous.
   e. Paleozoic.

7. How could the accumulation of carbon dioxide in the ocean lead to the extinction of some forms of animal life?
   a. Excess carbon dioxide would have killed many of the plant species, reducing the available supply of food for animals.
   b. Excess carbon dioxide would make it difficult for animals to metabolize.
   c. Excessive carbon dioxide would cause many fires that would kill many of the species of plants.
   d. Both a and b.
   e. Both b and d.

8. How does the fossil record support the hypothesis that an accumulation of carbon dioxide in ocean waters led to the mass extinctions that occurred in the Permian?
   a. High concentrations of carbon dioxide are present in fossils of the Permian.
   b. High concentrations of metal oxides are present in fossils of the Permian.
   c. Marine species that were intolerant of carbon dioxide were eliminated, while those with more tolerance for carbon dioxide survived.
   d. The excess carbon dioxide led to the extinction of plankton and ultimately other more advanced animal species.
   e. None of the above.

9. Which of the following appeared at about the same time as the dinosaurs?
   a. the insects
   b. the birds
   c. pterosaurs
   d. amphibians
   e. rodents

10. Not all scientists accept the hypothesis that the dinosaurs became distinct as the result of a large meteor strike. Evidence that supports the meteor hypothesis is
    a. the presence of radioactive material in fossils of dinosaurs that died at that time.
    b. that fossilized dinosaurs that died at that time obviously did so from a lack of oxygen.
    c. the large amounts of soot that were deposited in rocks worldwide at the time of dinosaur extinction.
    d. infectious disease could not have killed so many species of dinosaurs.
    e. both a and c.

11. Which of the following might be a reason that the modern fishes were more successful than the agnathans?
    a. The agnathans were only able to feed on foods found on the bottom of the ocean, while modern fishes can take advantage of more diverse food sources.
    b. Modern fishes were hunters and sought food in diverse places, while the agnathans were limited to filter feeding on the ocean's bottom.
    c. Modern fishes were more efficient at removing oxygen from the water.
    d. The agnathans were unable to produce the large number of offspring produced by modern fishes.
    e. None of the above.

12. A bony skeleton versus bony body plates in fish is an advantage because it
    a. supported strong muscles that enabled fish to chase prey.
    b. provided better protection from predators.
    c. allowed fish to grow larger and become more efficient predators.
    d. allowed fish to swim in the upper levels of the water.
    e. allowed fish to survive in freshwater habitats.

13. A swim bladder allows a fish to
    a. regulate its water content in seawater.
    b. regulate its salt content in seawater.
    c. regulate its salt content in freshwater.
    d. control its buoyancy in water.
    e. tolerate changes in water pressure.

14. The monotremes are classified as mammals because they
    a. are endoderms.
    b. are carnivores.
    c. are omnivores.
    d. have mammary glands.
    e. give birth to live young rather than lay eggs.

15. Fish resist dehydration in a marine environment by
    a. maintaining a higher internal concentration of salt than is in the water.
    b. reabsorbing water through their kidneys.
    c. secreting a slime layer to cover their body.
    d. both a and c.
    e. none of the above.

Matching

1. __________ Thecodonts
2. __________ Crocodiles
3. __________ Pelycosaurs
4. __________ Therapsids
5. __________ Pterosaurs

a. *Dimetrodon* is an example of these animals; they were the first land vertebrates able to kill prey their size
b. These large reptiles have changed very little since they first appeared
c. Flying dinosaurs
d. They were the first reptiles to stand on two feet
e. They were hungry reptiles that were the immediate ancestors of mammals

Completion

1. Vertebrates evolved in the oceans about __________ years ago and invaded land about __________ years ago.
2. The Mesozoic era lasted from about __________ to __________ million years ago and was a time of intensive evolution of __________ .
3. During the __________ period, __________% of marine animal species became extinct.
4. The Mesozoic is made of three periods: __________ , __________ , and __________ .
5. The climate of the early Cenozoic was __________ .
6. Large fish with heavy bony plates were called __________ .
7. Modern fish rely on the __________ to detect changes in water pressure.
8. The bony fish are members of class __________ , of which there are about __________ species.
9. Water moving past fish gills in the same direction as the fish is moving permits __________ , which is efficient for extracting oxygen from water.
10. Mammal lungs have many small chambers called __________ that are clustered together like grapes.
11. A significant difference between the heart of fishes and amphibians is that amphibians have a __________ .

Cool Places on the 'Net

To subscribe to the Food and Drug Administration mailing list send an e-mail message to:

http://www.fda.gov/

The National Library of Medicine can be accessed at:

http://www.nlm.nih.gov/

Label the Art

Label the parts of the egg.

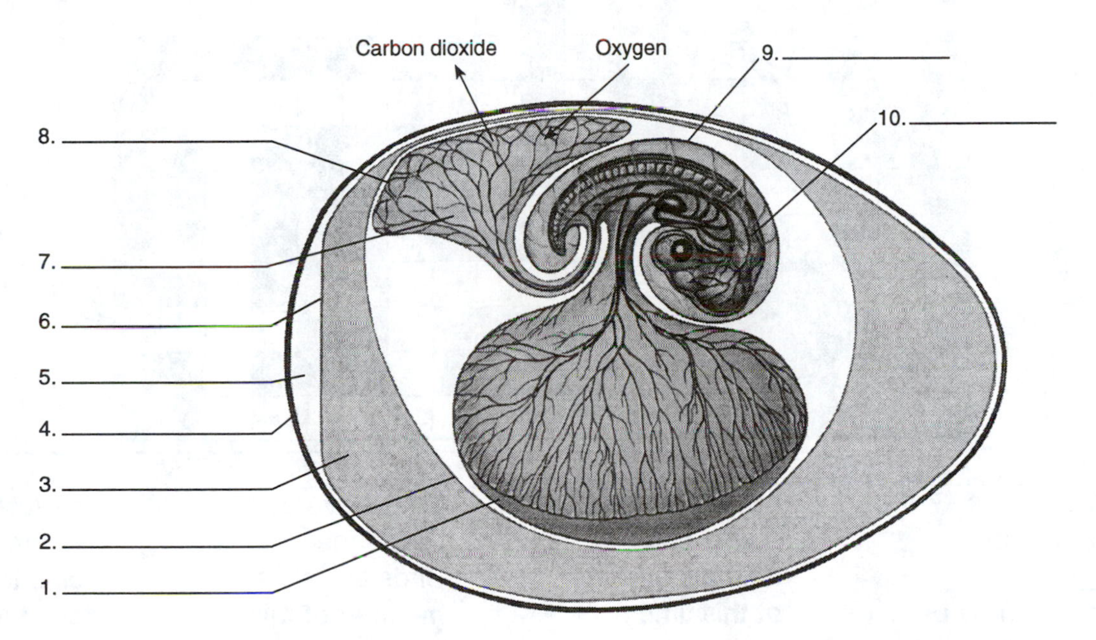

Crossword

ACROSS

2. Amphibians invaded the ______
4. Periods may be divided into these
6. They have not changed much
7. There were lots of dinosaurs at that time
12. Ancient wing
14. _____ extinctions
15. This came after the Cambrian period
17. Jawless fishes
19. Helpful for breathing out of water
21. Lots of land plants and animals evolved at this time
23. Modern fishes are this
24. Very old remains of an organism

DOWN

1. They have a backbone
3. Birds are a member of this class
5. Necessary to carry a developing embryo
8. They have a cartilage skeleton and a big appetite
9. It started about 65 M.Y.A.
10. Flying reptiles
11. Most species are found in Austrailia
13. They were the ancestors of horseshoe crabs
16. The lateral ____ system in fish helps them detect movement in water
18. Disposal units of the kidney
19. They evolved from fins
20. Fish use them to breathe
22. They are divided into periods

18 How Humans Evolved

Chapter Concepts

Primates are the order of mammals that contains humans. Prosimians were the first primates, small and nocturnal. Prosimians were replaced by their descendants, which are day-active. Apes also evolved from prosimians and have larger brains.

Two genera are considered hominid (of the human line): *Australopithecus* and *Homo.* Hominids are characterized by upright walking and large brains. Many stocky species of *Australopithecus* appear to be side branches on the evolutionary tree. The slender species like Lucy appear to be our immediate ancestors. The oldest hominids are 4.2 million years old.

The first human, *Homo habilis,* appeared 1.8 million years ago. The second species of human, *H. erectus,* appeared in Africa about 1.5 million years ago and survived for a million years, longer than any other species of human. *H. erectus* migrated out of Africa to Europe and Asia.

Our species, *Homo sapiens,* evolved in Africa about half a million years ago. Migrating out of Africa, *H. sapiens* retraced the spread of *H. erectus,* eventually supplanting it. Early *H. sapiens,* called Neanderthals, actually had bigger brains than Cro-Magnons (modern) *H. sapiens.*

Multiple Choice

1. The animal that is thought to be the ancestor to primates was
   a. a large, ground-dwelling carnivore.
   b. small and had a diet that consisted primarily of seeds.
   c. a marsupial that ate fruit.
   d. mouse-sized and ate insects.
   e. capable of flight over short distances.

2. Human dentition includes
   a. flattened molars.
   b. pointed canines.
   c. chisel-shaped incisors.
   d. all of the above.
   e. both a and c.

3. Which of the following has provided scientists with useful information about extinct animals?
   a. dentition
   b. fur
   c. size
   d. mating behavior
   e. burrows

4. Which of the following was a common animal about 38 million years ago?
   a. apes
   b. chimps
   c. prosimians
   d. dinosaurs
   e. early humans

5. Today most prosimians are found in (on)
   a. North America.
   b. parts of Australia.
   c. Europe.
   d. Asia.
   e. Madagascar.

6. The line of apes that led to gibbons diverged from other apes about __________ years ago.
   a. 50 million
   b. 50,000
   c. 25 million
   d. 25,000
   e. 10 million

7. A characteristic of hominoids is
   a. a larger brain.
   b. bipedalism.
   c. nighttime activity.
   d. monogamy.
   e. both a and b.

8. Evidence supporting the hypothesis that primates walked upright prior to the acquisition of large brains is provided by
   a. the leg structure of lemurs.
   b. the fossilized remains of Lucy.
   c. fossils of Java man.
   d. the structure of prosimian feet.
   e. none of the above.

9. Early European humans were called
   a. Java man.
   b. Neanderthals.
   c. Peking man.
   d. Cro-Magnon.
   e. none of the above.

Matching

1. __________ *Australopithecus robustus*
2. __________ *Australopithecus africanus*
3. __________ *Australopithecus afarensis*
4. __________ *Australopithecus ramidus*
5. __________ *Australopithecus boisei*

a. He was called "nutcracker man" due to his powerful jaws
b. Near apes that might be early ancestors to modern apes
c. Dart called him the "missing link"
d. Fossils found in 1938 in South Africa
e. Lucy is this genus and species

Completion

1. In 1871 Darwin published yet another controversial book called __________ .
2. In addition to binocular vision the primates also have __________ .
3. __________ are the surviving prosimians.
4. Human and chimpanzee DNA differs by about __________ .
5. __________ man fossils are about 500,000 years old.
6. Fossil evidence of Peking man shows that he used __________ and __________ .

Cool Places on the 'Net

The American Medical Association has a home page at:

http://www.ama-assn.org/

Key Word Search

S	U	T	C	E	R	E	O	M	O	H	Y	R	Z	K
K	S	N	E	I	P	A	S	O	M	O	H	M	H	K
C	H	C	Y	P	Y	L	U	L	V	M	M	L	O	N
L	O	M	A	R	U	M	C	G	L	O	A	J	U	G
S	M	P	B	N	S	V	U	Z	V	H	E	D	E	J
V	I	R	V	C	I	S	C	P	T	A	I	M	F	B
F	N	O	B	E	W	N	R	R	B	B	Y	C	D	E
B	O	S	Y	Z	Q	E	E	O	V	I	I	S	P	I
Y	I	I	C	K	M	D	Y	S	S	L	O	Y	U	T
Y	D	M	U	O	N	A	M	G	N	I	K	E	P	S
M	S	I	L	A	D	E	P	I	B	S	C	K	O	C
W	I	A	E	L	E	K	R	E	V	H	J	N	M	H
V	R	N	A	M	A	V	A	J	S	A	J	O	I	L
S	D	S	E	T	A	M	I	R	P	W	V	M	R	C
F	C	Z	H	B	D	R	W	T	R	X	I	Z	R	V

Apes
Bipedalism
Canines
Hominoids
Homo erectus
Homo habilis

Homo sapiens
Incisors
Java man
Lucy
Molars
Monkeys

Neanderthal
Peking man
Premolars
Primates
Prosimians
Prosimians

Crossword

ACROSS	DOWN

ACROSS

1. Australopithecus and humans are these
6. Sharp, pointed teeth
8. Monkeys first evolved in central ________
10. It means "before monkeys"
13. Flattened teeth used for grinding
14. The country where Lucy was found
15. Humans evolved from these primates
18. Homo erectus was also called ______ man
19. The line of animals that led to humans
20. She was "very together"

DOWN

1. Homo ______ was a "handy" guy
2. They were early European humans
3. He wrote "The Descent of Man"
4. Homo ______ means "wise man"
5. _____ man probably liked cafe latte
7. Sharp chisel-shaped teeth
9. It means "ape"
11. He worked on Leakey's project
12. They replaced the "before monkeys"
16. Human and chimp hemoglobin varies by this many amino acids
17. Walking on two feet

19 The Human Body

Chapter Concepts

The human body is organized like that of all vertebrates in general structure. Its 100 trillion cells are organized into tissues, which are the actual structural and functional units of the human body.

Adult tissues are grouped into four general classes: epithelium, connective tissue, muscle, and nerve. The outermost of the principal tissues, epithelium, protects the tissues beneath it from dehydration. Connective tissue supports the body structurally, defends it with the immune system, and transfers materials via the blood.

The skeleton protects the internal organs of the body and provides a strong and rigid base against which muscles can pull. Bone is a dynamic tissue, constantly growing and renewing itself.

Muscle cells do the actual work of movement. Muscle cells contain large amounts of the protein filaments actin and myosin. Muscle cells contract when myosin "walks" along the actin filament, driven by the cleavage of ATP.

Multiple Choice

1. Epithelial tissue arises from embryonic
   a. ectoderm.
   b. mesoderm.
   c. endoderm.
   d. neural tube cells.
   e. fluid.

2. Epithelial tissue functions to
   a. secrete a variety of materials.
   b. protect other tissues.
   c. provide sensory surfaces.
   d. all of the above.
   e. both a and b.

3. Hormones are secreted by what type of tissue?
   a. connective
   b. nervous
   c. muscle
   d. epithelial
   e. skeletal

4. Macrophages originate from which embryonic tissue?
   a. ectoderm
   b. mesoderm
   c. endoderm
   d. neural tube
   e. none of the above

5. What type of connective tissue is made of collagen fibers that are coated with calcium phosphate salt?
   a. adipose
   b. fibroblasts
   c. bone
   d. cartilage
   e. lymphocytes

6. __________ function(s) to attack cells that are infected with viruses.
   a. Adipose
   b. Fibroblasts
   c. Bone
   d. Cartilage
   e. Lymphocytes

7. Myofibrils are
   a. the basic unit of the muscle cell.
   b. microfilaments bunched into fibers.
   c. only found in skeletal muscle.
   d. only found in cardiac muscle.
   e. necessary for muscle contraction.

8. Glial cells
   a. transmit nerve impulses.
   b. span gaps between neurons.
   c. secrete neurotransmitters.
   d. supply neurons with nutrients.
   e. act as antennae for neurons.

9. The human body has a spine that is made up of __________ vertebrae. There are __________ pairs of ribs that curve forward from the vertebrae.
   a. 25, 10
   b. 33, 12
   c. 41, 10
   d. 50, 12
   e. 52, 10

10. Osteoblasts
    a. make up Haversian canals in bones.
    b. produce the bone marrow.
    c. secrete collagen.
    d. lose their nucleus upon reaching maturity.
    e. are found in reptiles but not in mammals.

Matching

1. __________ Erythrocytes
2. __________ Lymphocytes
3. __________ Plasma
4. __________ Fibroblasts
5. __________ Macrophages

a. Cells that engulf and digest invading microorganisms
b. Cells that transport oxygen and carbon dioxide in the body
c. Antibodies are produced by these cells
d. This is the fluid in which red blood cells are suspended
e. These cells secrete proteins into the spaces between cells

Completion

1. The four types of human tissue types are __________, __________, __________, and __________.
2. The human body contains about ________ different kinds of cells.
3. List the 11 principal organ systems in the human body.

   __________ __________
   __________ __________
   __________ __________
   __________ __________
   __________ __________
   ____

4. __________ are a connective tissue in which the individual cells are not attached to each other and that look like a flattened sphere with a depressed center.
5. Contractible protein fibers in muscle cells are called __________.
6. The human skeleton consists of __________ bones; __________ of these bones make up the axial skeleton, while __________ bones make up the appendicular skeleton.

Cool Places on the 'Net

The University of Pennsylvania has an online oncology (cancer) database called ONCOLINK. It can be accessed at:

http://cancer.med.upenn.edu/about_oncolink.html

Label the Art

Label the parts of the neuron.

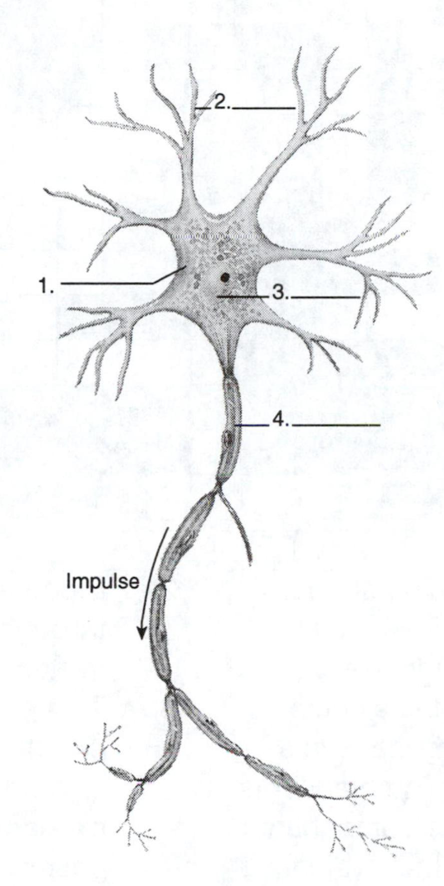

From John W. Hole, Jr., *Human Anatomy and Physiology,* 5th ed.

Crossword

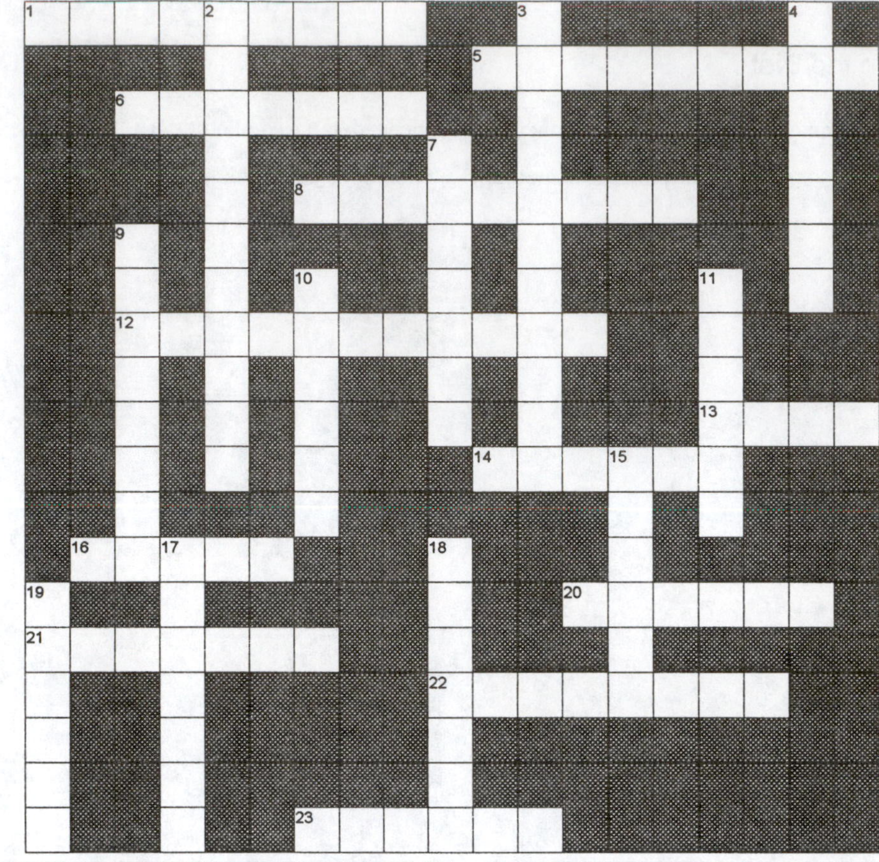

ACROSS		DOWN	

ACROSS

1. Glands and hormones
5. Bones that surround the spinal cord
6. This tissue is rather "excitable"
8. The _______ cavity contains the stomach, intestines and liver
12. They secrete collagen
13. It carries impulses away from the cell body
14. The longer fiber in muscle
16. The shorter filament in muscle
20. Less compact bone is this
21. The bladder is a component of this system
22. It's made up of bones
23. The fluid portion of blood

DOWN

2. Blood moves through this system
3. This system captures oxygen and exchanges gases
4. The heart muscle
7. The outer layer of bone is called this
9. This cavity contains the heart and lungs
10. Connections between bones
11. Different tissues grouped together
15. This type of epithelium is only one layer thick
17. They attach muscles to bones
18. Groups of cells of the same type
19. A type of tissue that moves things

20 Circulation and Respiration

Chapter Concepts

Blood is pumped out from the heart through arteries and returns through veins. Materials pass in and out of the blood as it passes through a network of tiny tubes called capillaries that connect arterial to veinous circulation. Fluid that passes out of the capillaries is recovered by the lymphatic system.

Blood is composed of a fluid called plasma, white blood cells of the immune system, and red blood cells that carry oxygen. Dissolved in plasma are many kinds of molecules, including a considerable amount of serum albumin protein that osmotically balances the blood.

The heart is a double pump. The right side pumps oxygen-poor blood to the lungs and then returns it via pulmonary veins to the left side. The left side pumps oxygen-rich blood to the body and then returns it to the right side. The contraction of the heart starts in the upper wall of the right atrium and spreads as a wave across the heart.

Respiration is the uptake of oxygen gas from air and the simultaneous release of carbon dioxide. Expansion of the chest cavity draws air into the lungs; relaxation forces it back out. CO_2 is transported largely as bicarbonate ion.

Cancer is unrestrained cell proliferation caused by damage to growth-regulating genes. Cigarette smoking is the principal cause of lung cancer. Smoking produces lung cancer by introducing carcinogens into the lungs.

Multiple Choice

1. Blood leaves the heart through
   a. veins.
   b. capillaries.
   c. arteries.
   d. venules.
   e. arterioles.

2. Blood moves from capillaries into
   a. veins.
   b. subcapillaries.
   c. arteries.
   d. venules.
   e. arterioles.

3. __________ carry blood back to the heart.
   a. Veins
   b. Capillaries
   c. Arteries
   d. Venules
   e. Arterioles

4. An important characteristic of an artery is
   a. flexibility.
   b. its radius of about 8 micrometers.
   c. its ability to return blood back to the heart.
   d. its affinity for carbon monoxide.
   e. both a and b.

5. Arterioles differ from arteries in that
   a. they are smaller in diameter.
   b. they are larger in diameter.
   c. hormones can cause the muscle surrounding them to relax and they can enlarge in diameter.
   d. both a and c.
   e. both b and c.

6. The thin walls of capillaries are important because they
   a. allow capillaries to easily change diameter.
   b. allow for the transport of gases and metabolites.
   c. must be flexible to accommodate surges of blood.
   d. must withstand high pressure.
   e. none of the above.

7. The lymphatic system functions to
   a. filter bacteria and debris from lymph fluid.
   b. remove carbon dioxide from the blood.
   c. add oxygen to the blood.
   d. adjust the water content of plasma.
   e. add bicarbonate to the blood.

8. The lymphatic system
   a. transports fats absorbed from the intestine.
   b. returns proteins to circulation.
   c. filters debris from lymph fluid.
   d. destroys debris removed from lymph fluid.
   e. all of the above.

9. A function of proteins dissolved in blood plasma is to
   a. provide energy to blood cells.
   b. prevent bacteria from surviving in the bloodstream.
   c. prevent the osmotic loss of water from the plasma.
   d. prevent salts from building up in the blood.
   e. maintain a constant pH in the blood.

10. Heart contraction is initiated by
    a. the superior vena cava.
    b. Purkinje fibers.
    c. an electrical impulse of the appropriate intensity.
    d. sinoatrial node.
    e. the bundle of His.

11. During a routine examination of your elderly dog, the veterinarian tells you that she has a slight heart murmur. A likely cause of the murmur is
    a. the heart valves encountering fat accumulation when they close.
    b. turbulence in the heart caused by valves opening or closing incompletely.
    c. failure of the heart to completely empty.
    d. weakening of the cardiac muscle.
    e. both a and d.

12. Alveoli are connected to the bronchi by
    a. blood vessels.
    b. connective tissue.
    c. muscle fibers.
    d. the pleural membrane.
    e. bronchioles.

13. The effect of carbon dioxide on hemoglobin is that
    a. it causes the hemoglobin to change shape and unload the oxygen it is carrying.
    b. it changes its shape so that hemoglobin can no longer bind to oxygen.
    c. it increases hemoglobin's affinity for oxygen.
    d. hemoglobin is denatured.
    e. both a and d.

14. Hemoglobin can bind to and transport
    a. oxygen.
    b. carbon dioxide.
    c. nitric oxide.
    d. all of the above.
    e. both a and b.

15. In order for lung cancer to be initiated
    a. several thousand genes must be mutated.
    b. only a few critical genes need to be mutated.
    c. lung cells must be exposed to some kind of radiation.
    d. the patient must be in poor health.
    e. none of the above.

Matching

1. __________ Megakaryocytes
2. __________ Hematocrit
3. __________ Erythrocytes
4. __________ Leukocytes
5. __________ Platelets

a. Bits of cytoplasm pinch off of these cells
b. They are filled with hemoglobin
c. Fibrin fibers cause these cells to stick together and assist in clotting
d. These cells can migrate out of the bloodstream
e. This refers to the volume of blood that is occupied by cells

Completion

1. The three components of the circulatory system are __________ , __________ , and __________ .
2. __________ , __________ , and __________ are found dissolved in blood plasma.
3. __________ is a condition in which fats accumulate in the arteries.
4. Cancerous growths that form a ball of cells are called __________ , while those that involve connective tissue are called __________ .

Cool Places on the 'Net

For information about heart disease contact the American Heart Association at:

http://www.amhrt.org/ahawho.htm

Label the Art

A. Label the parts of the human cardiovascular system.

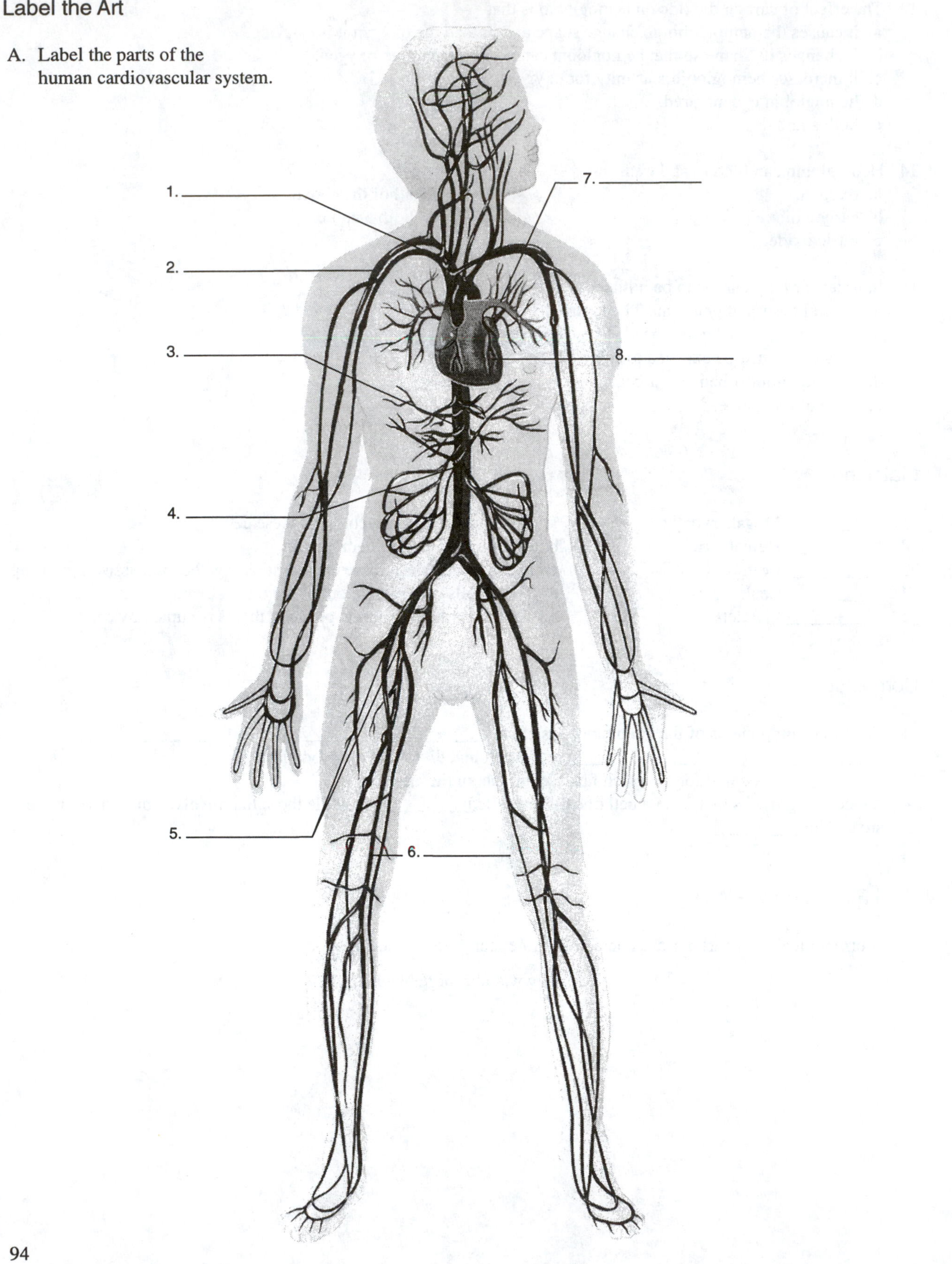

B. Label the parts of the human respiratory system.

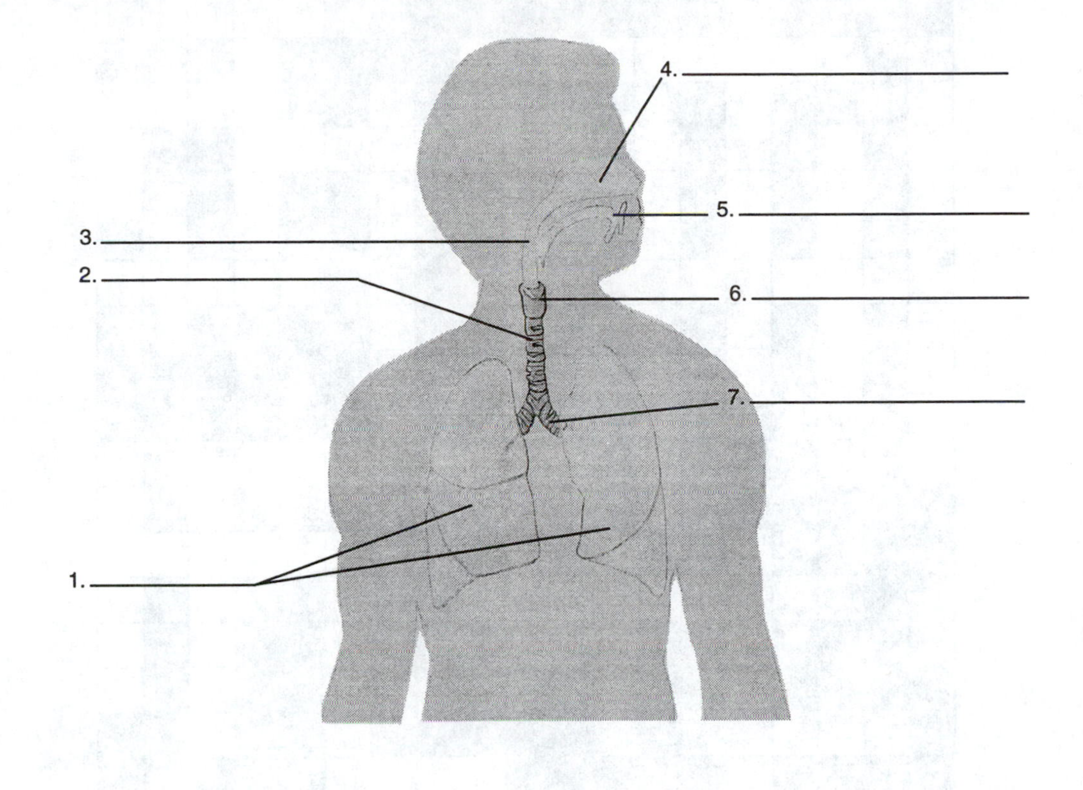

Crossword

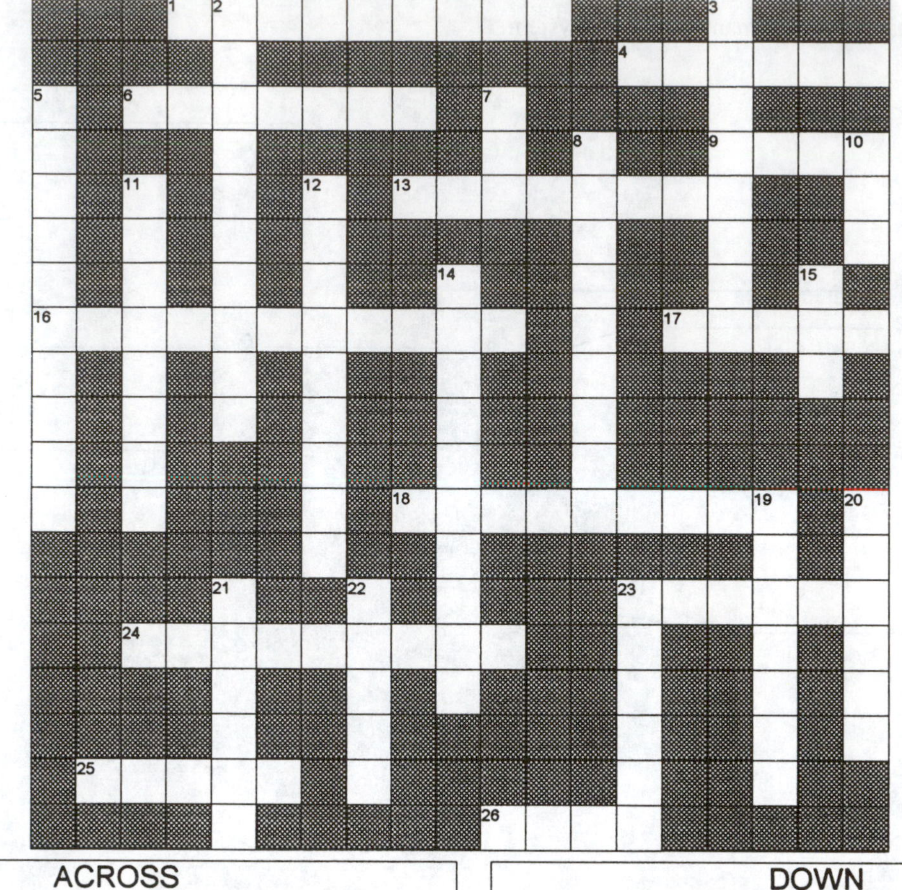

ACROSS

1. Epithelial tissue tumor
4. Vertebrates have this kind of circulatory system
6. It can cause cancer
9. It fills when you breathe
13. Lymph nodes help remove these from the circulatory system
16. Where food and oygen molcules are transferred from
17. _____ return blood to the heart
18. They assist in the clotting process
23. This valve prevents blood from reentering the atrium
24. Low blood pressure
25. _____ pressure
26. Carbon dioxide has this effect on oxygen unloading

DOWN

2. They have a smaller diameter than arteries
3. When empty, veins do this
5. Another name for white blood cells
7. Short for red blood cell
8. Blood passes through the left _______ after leaving the left atrium
10. _____ exchange
11. A thick layer of muscle lining the bottom of the thoracic cavity
12. Moving air in and out of the lungs
14. A molecule that carries oxygen
15. The average capillary is about _____ millimeter long
19. Connective tissue tumor
20. The air left when a breath is complete
21. _____ dioxide is removed by hemoglobin
22. It is a large blood vessel
23. A heart "noise"

21 The Path of Food Through the Body

Chapter Concepts

Ingested calories are either metabolized by the body or stored as fat. Healthy diets contain those necessary substances we cannot manufacture such as essential amino acids, vitamins, and trace elements.

Digestion is the breaking down of macromolecules into small metabolizable components. Digestion is carried out by enzymes. Protein digestion is aided by acid, which opens up folded proteins so that enzymes can attack them. Acid predigestion occurs in the stomach; most enzymatic digestion occurs in the duodenum, the initial portion of the small intestine.

The liver, the body's largest internal organ, monitors metabolism. The liver regulates levels of glucose in the blood by storing excess glucose as glycogen and then converting it to fat. The hormone insulin stimulates this conversion of glucose to glycogen. When blood glucose levels fall, another hormone called glucagon stimulates the breakdown of glycogen to glucose.

Kidneys cleanse the blood of many harmful substances. Among the most important of these are the nitrogen by-products of protein metabolism.

A major function of the kidneys is water conservation. A second key function of the kidneys is to regulate how much salt the blood contains. Mammals concentrate their urine by removing water from it; this allows them to retain the water when the urine is discarded.

Multiple Choice

1. Fats have a higher energy content per gram than do carbohydrates because they
   a. are more easily converted into ATP.
   b. are more easily metabolized.
   c. have more energy-rich bonds.
   d. release their energy at a lower temperature.
   e. none of the above.

2. Fats are a necessary part of the diet to
   a. provide energy.
   b. manufacture cell membranes.
   c. insulate nervous tissues.
   d. build a variety of structures in the cell.
   e. all of the above.

3. A diet that is high in fiber is important because
   a. the bulk helps to move food through the colon at a faster rate.
   b. fiber is a source of essential amino acids.
   c. fiber is an important source of nutrients.
   d. fiber is an important source of vitamins.
   e. the majority of protein in foods is in the bulk.

4. A consequence of a diet lacking vitamin C is
   a. the inability to synthesize hemoglobin.
   b. the development of diabetes.
   c. the development of scurvy.
   d. memory loss.
   e. the inability to breakdown carbohydrates.

5. Trace elements are
   a. only necessary for plant growth.
   b. obtained from the plants that we consume in our diet.
   c. obtained from the animal products in our diet.
   d. both a and b.
   e. both b and c.

6. __________ is(are) necessary to break up proteins, lipids and other nutrients to release energy.
   a. High temperatures
   b. Acids
   c. Enzymes
   d. Trace elements
   e. Oxygen

7. Chyme is
   a. the enzyme and acid mixture that breaks down food in the stomach.
   b. the name used for the infoldings of epithelial tissue inside of the stomach.
   c. the name of the sphincter located at the end of the esophagus.
   d. a mixture of digestive juices and food.
   e. the mucosa that lines the inside of the stomach and protects it from the acids located there.

8. Gastrin functions to
   a. enhance the production of HCl in the stomach.
   b. regulate the synthesis of HCl in the stomach.
   c. absorb fats in the small intestine.
   d. regulate water loss from the large intestine.
   e. kill any bacteria that enter the stomach.

9. Aspirin can cause an upset stomach because it
   a. increases acid production.
   b. decreases acid production and slows digestion.
   c. is absorbed through the lining of the stomach.
   d. slows the breakdown of fats.
   e. increases bacterial multiplication in the stomach.

10. Gastric pits
    a. are lesions in the mucosa of the stomach.
    b. invaginations in the stomach epithelium.
    c. are the site of actual digestion in the stomach.
    d. in the small intestine produce digestive enzymes.
    e. are not acquired until middle age and are due to a diet low in fiber.

Matching

1. __________ Small intestine
2. __________ Stomach
3. __________ Duodenum
4. __________ Large intestine
5. __________ Liver

a. Sodium and vitamin K are absorbed through this structure, and it acts as a refuse dump for undigested food
b. Bicarbonate and digestive enzymes are delivered to this site
c. Bile salts are produced by this structure
d. Villi and microvilli increase the absorptive surface of this site
e. Gastrin is produced here

Completion

1. On the average, fats contain __________ calories per gram, while carbohydrates have __________ calories per gram.
2. Digestion takes place in two places: the __________ and the __________ .
3. In the digestive process, the tongue functions to __________ .
4. The tube that connects the mouth to the stomach is the __________ .
5. Digestive juices located in the stomach are produced by the __________ .
6. Blood sugar levels are regulated by __________ .

Cool Places on the 'Net

To find nutrition information on MedWeb go to:

http://www.gen.emory.edu/medweb/medweb.nutrition.html

FDA's web server for nutrition documents can be located at:

http://www.fda.gov/search.html

Label the Art

A. Label the parts of the human digestive system.

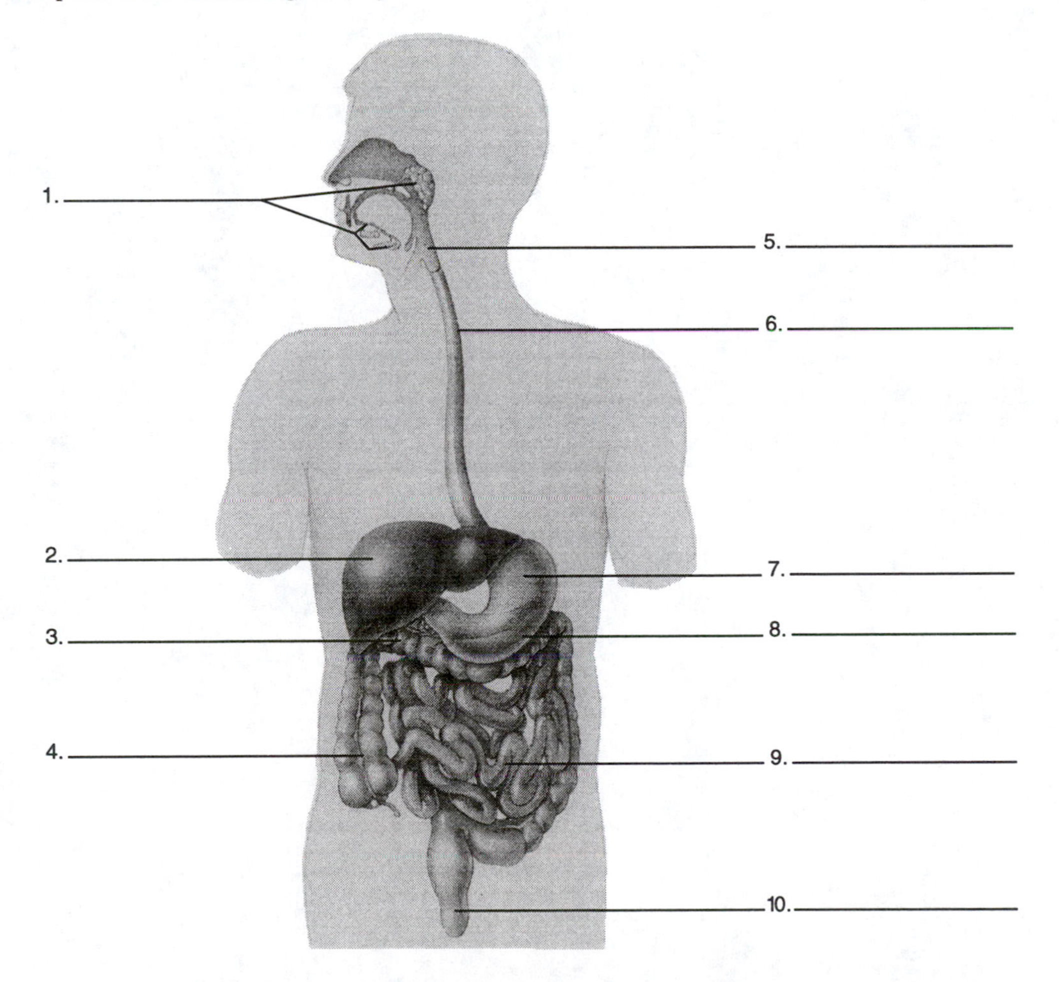

B. Label the parts of the human tooth.

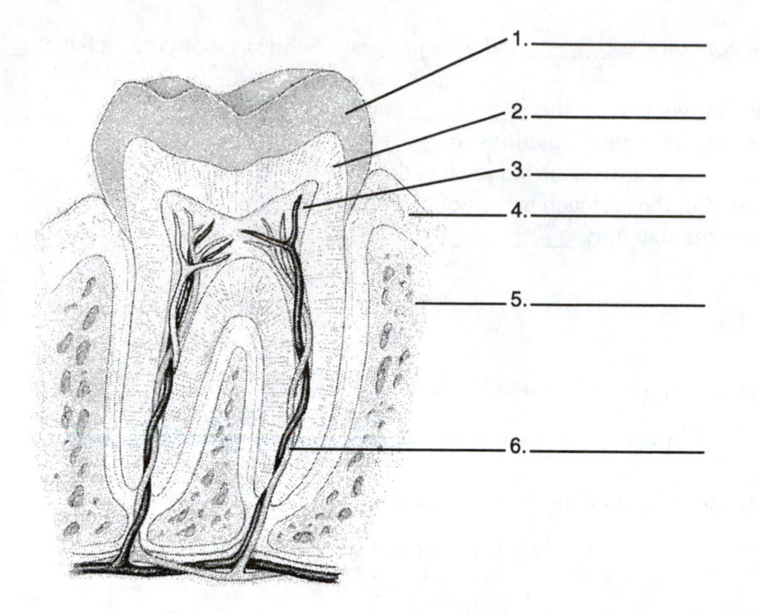

Crossword

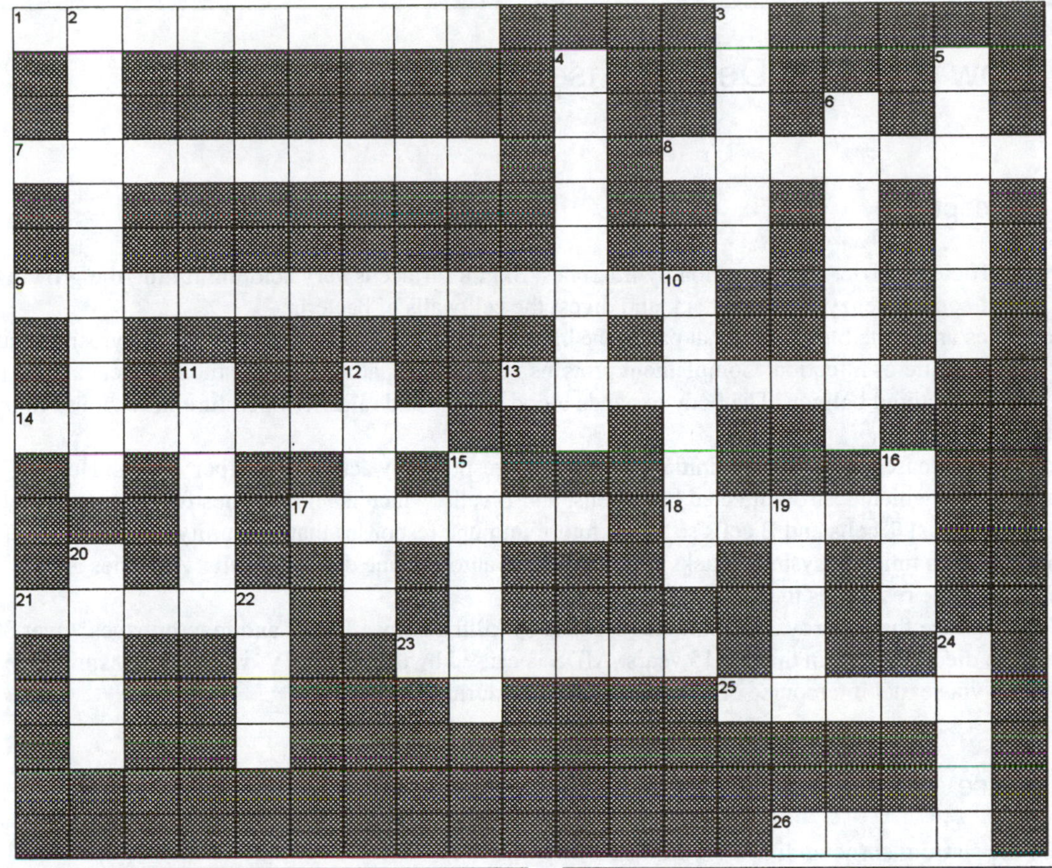

ACROSS

1. A muscular ring
7. The breakdown of foods is accomplished by this process
8. An enzyme that breaks down starch
9. Your body uses them for energy
13. These are often counted
14. C and E are just two of these
17. The whole ones are especially healthy
18. No digestion occurs here, and about 4% of water absorption does occur here
21. Eat 6 to 11 servings of ______ and cereals
23. It is affected by diet and exercise
25. ______ elements are required in small amounts
26. These should be used sparingly in the diet

DOWN

2. These contractions move food through the esophagus
3. He had a capsule
4. ______ amino acids must be included in our diet
5. Hormone that regulates synthesis of HCl
6. Starchlike chain of glucose subunits
10. A saclike structure that temporarily stores food
11. Moistens and lubricates food
12. The human body requires this to move and grow
15. Eliminates nitrogenous wastes
16. The end of the line
19. Bile salts are produced by this
20. The end result after salts, nutrients, and water have been eliminated
22. A four-letter word, for sure
24. Undigested material from the digestive system

22 How the Body Defends Itself

Chapter Concepts

Skin offers an efficient barrier to penetration by microbes. Skin's surface is very acid, inhibiting the growth of microbes. Sweat contains enzymes that attack and digest the cell walls of bacteria.

Macrophages are white blood cells that patrol the bloodstream, ingesting bacteria by phagocytosis. Neutrophils kill all the cells at a site of infection. Complement proteins insert into a pathogen's cell membrane, causing the cell to rupture like a punctured balloon. The body expands blood vessels and raises temperature in response to infection to aid the body's defenses.

Macrophages release chemicals that initiate the immune response by activating helper T cells. Helper T cells in turn activate T cells, which destroy infected body cells, and B cells, which label microbes for destruction by macrophages. Residual T cells and B cells so speed future immune responses that immunity results.

When the body's immune system attacks its own tissues, autoimmune disease results. Allergies are inappropriate immune responses to harmless substances.

AIDS is a disease that destroys the immune response by killing helper T cells and macrophages. Over 300,000 Americans have died of AIDS in the last 15 years. AIDS is caused by the virus HIV, which is transmitted in body fluids, typically via sexual intercourse or by infected needles during drug use.

Multiple Choice

1. Cells in the stratum corneum live there for about
   a. 8 hours.
   b. a day.
   c. a week.
   d. a month.
   e. 2 to 3 months.

2. A membrane attack complex is
   a. a group of macrophages that are attacking an invading bacterium or virus.
   b. a group of activated complement molecules.
   c. necessary for mast cells to release histamines.
   d. found in viral infections only.
   e. formed when B cells secrete antibody molecules.

3. Complement is activated by
   a. exposure to proteins released by viral infected cells.
   b. the lysis of neutrophils.
   c. encounter with bacterial or fungal cell walls.
   d. a decrease in vitamin C levels.
   e. both a and d.

4. The redness and swelling that are associated with inflammation are due to
   a. histamines.
   b. the action of neutrophils.
   c. chemical signals released from damaged cells.
   d. the production of interleukins by macrophages.
   e. waste products released by natural killer cells.

5. Macrophages
   a. phagocytize invading microbes in the body.
   b. produce interleukins.
   c. identify foreign cells in cooperation with B and T cells.
   d. examine the major histocompatibility proteins on the surface of the body's cells.
   e. all of the above.

6. Which of the following is the best description of the relationship between helper T cells and interleukin-1?
   a. Helper T cells secrete interleukin-1 when they encounter a viral infected cell.
   b. Interleukin-1 is a surface protein found on helper T cells.
   c. Interleukin-1 is secreted by natural killer cells and it activates helper T cells.
   d. Macrophages secrete interleukin-1, which then activates helper T cells.
   e. Helper T cells secrete interleukin-1, which then activates macrophages.

7. Killer T cells help to eliminate infected body cells and cancer cells by
   a. secreting antibodies to mark the cells for destruction.
   b. phagocytosis.
   c. fusing with them and secreting histamines that cause them to lyse.
   d. puncturing the surface of the cell.
   e. both a and d.

8. The production of antibodies is initiated by the
   a. release of interleukin-1.
   b. release of interleukin-2.
   c. release of histamines.
   d. development of a fever.
   e. killer T cells.

9. Mutations in the *nef* gene
   a. render B cells ineffective.
   b. result in increased allergies.
   c. inhibit the release of histamines.
   d. render HIV nonvirulent.
   e. affect the ability of macrophages to release interleukin-1.

10. The symptoms of lupus are caused by
    a. destruction of the myelin on motor nerves by the immune system.
    b. an attack on the tissues of the joint by the immune system.
    c. an attack on the connective tissue and kidneys by the immune system.
    d. an attack on the thyroid by the immune system.
    e. destruction of pancreas cells by the immune system.

11. Multiple sclerosis is caused by
    a. destruction of the myelin on motor nerves by the immune system.
    b. an attack on the tissues of the joint by the immune system.
    c. an attack on the connective tissue and kidneys by the immune system.
    d. an attack on the thyroid by the immune system.
    e. destruction of pancreas cells by the immune system.

12. Graves disease is caused by
    a. destruction of the myelin on motor nerves by the immune system.
    b. an attack on the tissues of the joint by the immune system.
    c. an attack on the connective tissue and kidneys by the immune system.
    d. an attack on the thyroid by the immune system.
    e. destruction of pancreas cells by the immune system.

13. HIV is transmitted by
    a. unprotected sex.
    b. protected sex.
    c. exposure to contaminated needles.
    d. all of the above.
    e. both a and c.

Matching

1. __________ Helper T cells
2. __________ Mast cells
3. __________ Macrophages
4. __________ Natural killer cells
5. __________ Neutrophils
6. __________ B cells

a. These cells are efficient at detecting cells that are infected by viruses
b. These cells help coordinate the activities of other immune cells
c. These cells release substances that are like household bleach to kill bacteria
d. A large number of these hungry cells are found in the spleen
e. Receptor molecules called antibodies are produced by these cells
f. These cells release histamines and other substances in allergic reactions

Completion

1. The body's first line of defense against infection is the __________ .
2. The three layers that make up the skin are __________ , __________ , and __________ .
3. The __________ involves redness and swelling due to increased capillary permeability.
4. Interleukin-1 is produced by __________ and influences __________ .
5. Receptor proteins secreted by B cells are called __________ .
6. When a gene from a pathogenic virus is inserted into another harmless virus, the result is a(n) __________ .
7. Diseases that result from the immune system attacking the body's cells are called __________ .

Cool Places on the 'Net

Medweb Immunology provides information about immunology and links to other related pages. Their address is:

http://www.cc.emory.edu/WHSCL/medweb.immunology.html

The World-Wide Web Virtual Library: Immunology is a great source of information about immunology research. Their address is:

http://golgi.harvard.edu/biopages/immuno.html

Label the Art

Label the parts of a section of human skin.

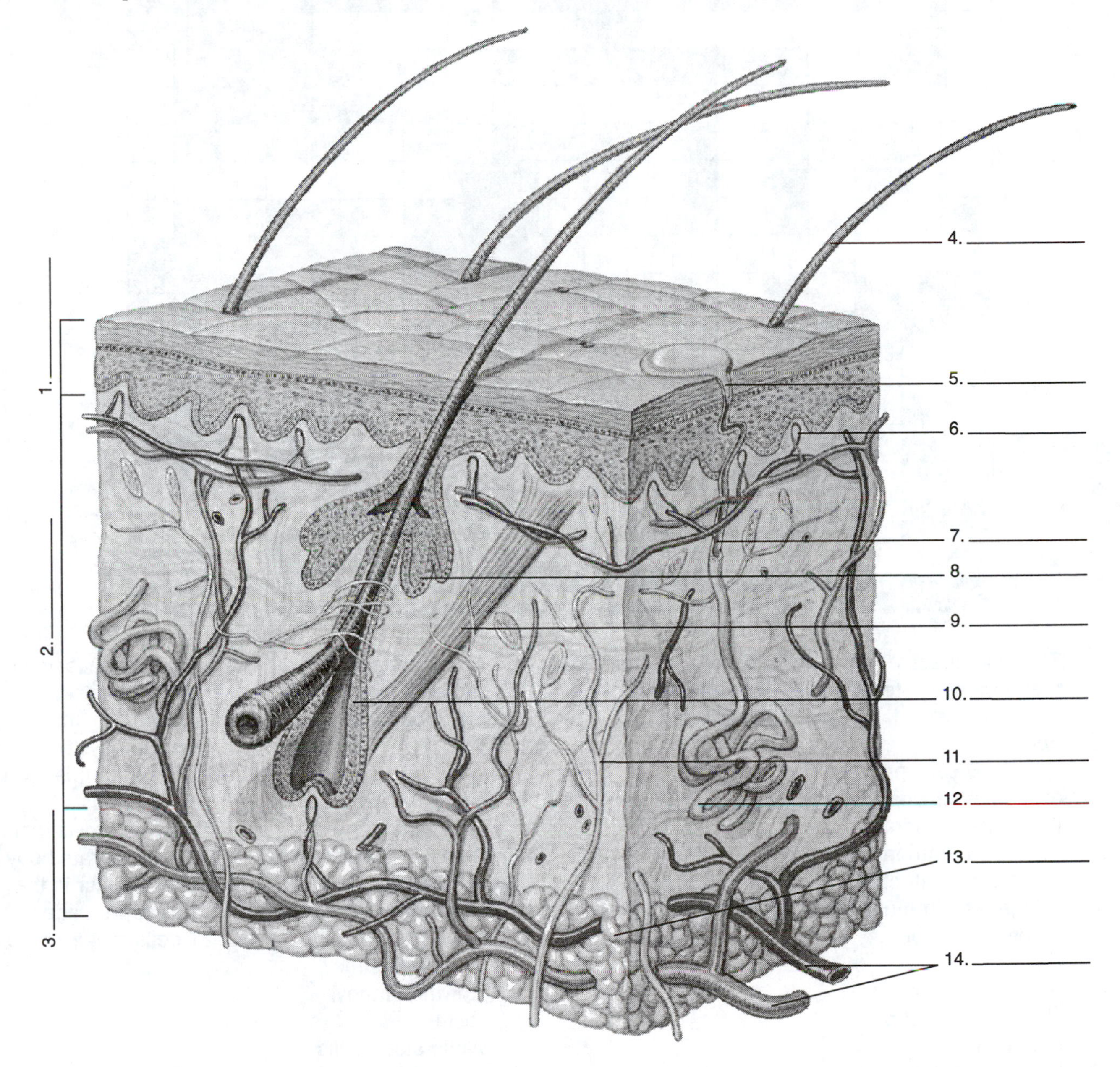

Crossword

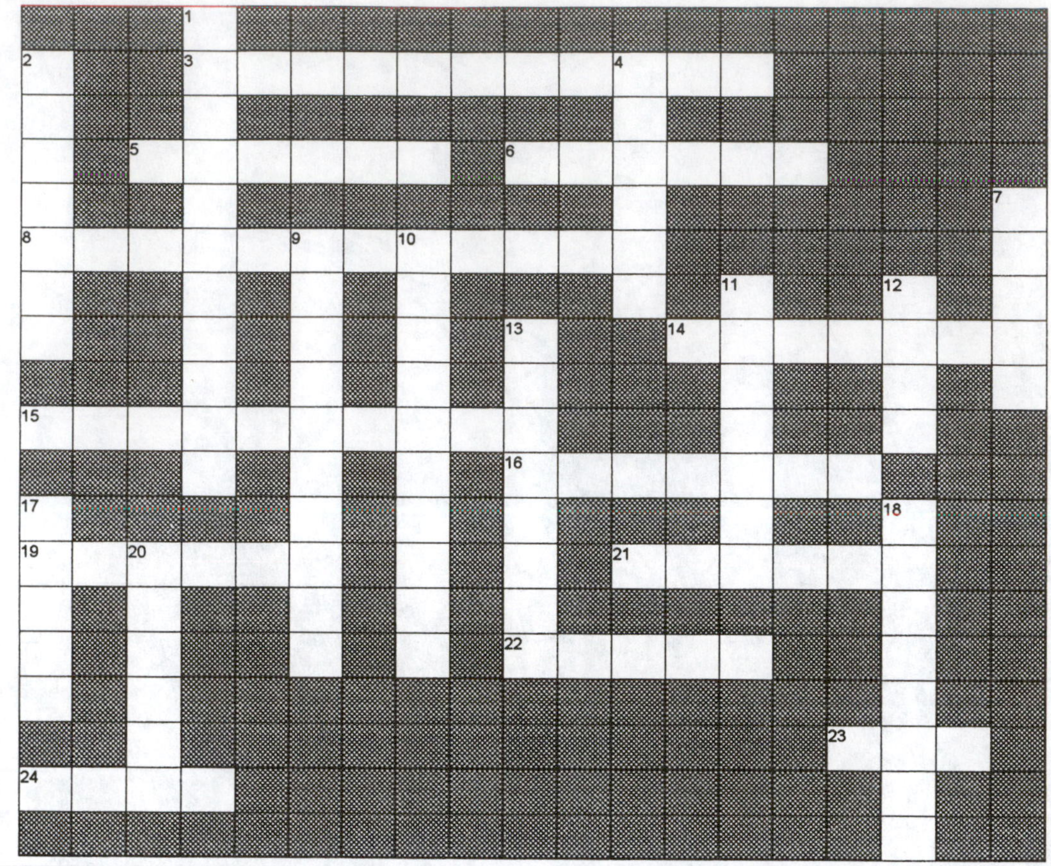

ACROSS

3. The "Kamikazes" of the immune system
5. It lies below the epidermis
6. ______ B cells hang around even when the battle is over
8. This is a common response to infection
14. Triggers an immune response without the disease
15. Protein markers
16. Hay fever is but one example
19. It sure makes it tough to breathe
21. B cell and T cell activity is controlled by this king of T cells
22. When you get hot you do this
23. A common disease caused by a virus
24. These cells release those nasty histamines

DOWN

1. An "alarm system" secreted by macrophages
2. The ______ of the skin discourages the growth of microbes
4. Our ______ system protects us against invading microorganisms
7. This inhibits the growth of many bacteria
9. White blood cells that ingest invading microbes
10. This kind of disease occurs when the body can't distinguish between "self" and "nonself"
11. ______ killer cells attack infected cells of the body
12. Disease caused by HIV
13. HIV ______ the immune system
17. Cells from this layer are the most actively growing of the vertebrate body
18. These occur in the dermis with age
20. T cells mature here

23 The Nervous System

Chapter Concepts

The nervous system is composed of sensory nerves that carry signals to the central nervous system and motor nerves that carry signals away from it.

The central nervous system is composed of the brain and spinal cord. Most of the mass of the brain is the cerebrum; its surface is the site of conscious thought. Other parts of the brain integrate information and coordinate body activities.

Voluntary motor nerves send commands from the brain to skeletal muscles. Involuntary motor nerves constitute the autonomic nervous system. The autonomic nervous system is composed of two antagonistic elements; the balance of the two determines the degree of stimulation.

Nerves are composed of the axons of many neurons, bundled together. Neurons expend ATP to pump out Na^+ ions and so polarize the cell membrane, generating a resting potential. A nerve impulse is a propagating depolarization of the cell membrane. Nerve impulses cross synapses via chemicals called neurotransmitters.

Interoceptors of many kinds monitor the body's internal condition. Sensing the exterior world focuses on three sorts of stimuli: chemicals, sound, and light. Both sound and light can be used to form three-dimensional images.

Multiple Choice

1. The autonomic nervous system
   a. controls the skeletal muscles.
   b. is active only when the body is in crisis.
   c. stimulates glands and controls the smooth muscles.
   d. controls all of the body's functions.
   e. none of the above.

2. A stroke is caused by
   a. blockage of the heart by clots.
   b. buildup of cholesterol in blood vessels.
   c. blockage of blood vessels in the brain by clots.
   d. insufficient thyroid hormone.
   e. lack of oxygen.

3. The reticular formation
   a. is only present in the brains of epileptics.
   b. is a network of nerves that runs through the brain connecting its parts.
   c. is sometimes called the medulla oblongata.
   d. is responsible for the control of smooth muscles.
   e. allows birds to maintain their balance as they fly.

4. The outer edges of the spinal cord are white because
   a. no red blood cells are present in the area.
   b. it is covered in a thick layer of connective tissue.
   c. of the presence of axons and dendrites.
   d. a layer of fat surrounds the structure.
   e. both a and c.

5. Interneurons are
   a. connecting neurons that are between the sensory and motor neurons.
   b. the spaces between dendrites and other neurons.
   c. stimulated by the pituitary gland.
   d. found on the epidermis.
   e. none of the above.

6. Interoreceptors
   a. regulate activity between adjacent neurons.
   b. detect information about the body's internal condition.
   c. work with neurotransmitters to regulate the speed of nerve impulses.
   d. open and close ion channels on sensory neurons.
   e. work only in the central nervous system.

7. A person with defective otolith sensory receptors
   a. has a difficult time maintaining balance.
   b. is deaf.
   c. cannot detect external temperature changes.
   d. has a faulty sense of smell.
   e. has a limited imagination.

8. Parallax is
   a. the synchronized functioning of neurons.
   b. a slight displacement of images that plays a role in distance perception.
   c. the path that light takes through the eye.
   d. only important in land animals.
   e. a defect caused by eyes that are oblong in shape.

Matching—Part 1

1. __________ Spinal cord
2. __________ Cerebrum
3. __________ Hypothalamus
4. __________ Cerebellum
5. __________ Medulla oblongata

a. Balance, posture, and muscular coordination are controlled here
b. Body temperature and blood pressure are controlled here
c. It is sometimes called the brain stem
d. This structure consists of sensory and motor nerve tracts
e. The "center for higher thought"

Matching—Part 2

1. __________ Neuroglial cells
2. __________ Myelin sheath
3. __________ Nodes of Ranvier
4. __________ Dendrite
5. __________ Schwann cells

a. Nerve impulses along the axon are facilitated by them
b. These extend from one end of a cell body
c. They make up more than half the volume of the human nervous system
d. A layer of insulation surrounding the axon
e. Gaps along the length of an axon

Completion

1. The __________ controls the pituitary gland and is linked to the __________ by a network of neurons.
2. The "fight-or-flight" reaction is controlled by the __________ .
3. The membrane located on the axon side of a synapse is called the ___________ , while the membrane on the receiving side of the synapse is the __________ .
4. In an inhibitory synapse, the receptor protein is a __________ .
5. Bats and shrews use __________ to move about in the dark.
6. Light entering the human eye first passes through the __________ .

Cool Places on the 'Net

The Glaxo Neurological Centre provides nonmedical advice and information about neurological illnesses. They can be found at:

http://www.connect.org.uk/merseyworld/glaxo/

The World-Wide Web Virtual Library: Neuroscience is a good starting point for information on neuroscience. Their address is:

http://neuro.med.cornell.edu/VL/

Label the Art

Label the parts of the human brain.

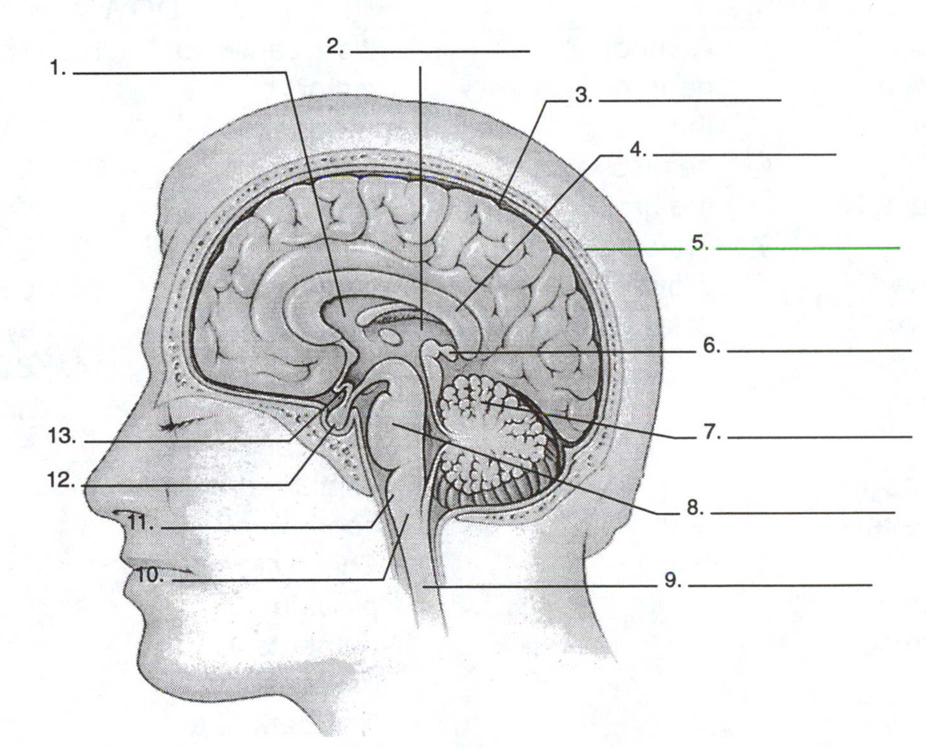

Crossword

ACROSS

2. It makes up about 85% of the weight of the brain
8. A membrane in the ear
9. They sense information about the body's internal condition
11. Some drugs have this effect
14. Bats and shrews have this in common
15. ________ nerves carry information towards the CNS
16. A bundle of neurons linking the hemispheres of the brain
17. These cells provide us with color vision

DOWN

1. It's a center for motor coordination
2. It resembles a snail
3. A light-detecting complex
4. ________ oblongata
5. This cannot be controlled by conscious
6. These cells provide nutrients to neurons
7. The bark of a tree
10. They envelope the axon with myelin
12. It is related to distance vision
13. Short for central nervous system
14. Between neurons

24 Chemical Signaling Within the Body

Chapter Concepts

Many of the hormone-producing endocrine glands are under the direct control of the central nervous system. The hypothalamus produces releasing hormones, which cause the pituitary to release particular hormones that circulate via the bloodstream to their target tissues.

Steroid hormones enter cells and interact with receptors located in the cytoplasm or nucleus. Peptide hormones interact with receptors located on the cell surface. Hormones often initiate a cascading series of events in the cell that greatly increases the power of the signal.

The anterior pituitary gland produces seven so-called pituitary hormones. Each of these hormones is released by a particular releasing hormone from the hypothalamus and circulates to a particular target tissue.

Multiple Choice

1. Hormones
   a. are relatively unstable and work only in the area adjacent to the gland that produced them.
   b. are stable, long-lasting chemicals released from glands.
   c. typically act at a site that is distant from where it was produced.
   d. are all lipid-soluble.
   e. both b and c.

2. Glands that are enclosed by tissue are called
   a. internal glands.
   b. protected glands.
   c. endocrine glands.
   d. pseudoglands.
   e. none of the above.

3. The central nervous system issues commands to the body's organs by means of
   a. the motor nervous system.
   b. the endocrine system.
   c. electrical impulses only.
   d. both a and b.
   e. none of the above.

4. How can a target cell recognize a particular hormone and not respond to other hormones?
   a. Target cells respond to whatever hormone is present in the largest concentration.
   b. The target cells are always located close to the source of the hormone, making it easy to respond to that hormone.
   c. Protein receptors located on the surface of the target cell or in the cytoplasm match the hormone.
   d. Carbohydrate tags on the surface of the target cell match the hormone.
   e. The hormone is only able to enter membrane channels on the correct target cell.

5. An example of a steroid hormone that influences secondary sexual characteristics is
   a. cortisone.
   b. prolactin.
   c. follicle-stimulating hormone.
   d. cyclic AMP.
   e. testosterone.

6. The steroid hormones bind to
   a. protein receptors in the cytoplasm of the target cell.
   b. protein receptors on the surface of the target cell.
   c. carbohydrate receptors in the cytoplasm of the target cell.
   d. carbohydrate receptors on the surface of the target cells.
   e. both a and c.

7. Low blood calcium levels cause the parathyroid gland to
   a. increase in size to form a goiter.
   b. stimulate the release of calcium from the bones.
   c. release insulin.
   d. produce additional quantities of PTH.
   e. none of the above.

8. Individuals with diabetes mellitus
   a. are unable to take up glucose from the blood.
   b. are unable to break down glucose.
   c. require additional insulin to break down glucose.
   d. are twice as likely as nondiabetics to have a heart attack.
   e. cannot consume glucose in their diet.

9. Individuals with type I diabetes
   a. have an autoimmune disease in which the islets of Langerhans are attacked.
   b. are allergic to insulin.
   c. have an abnormally low number of insulin receptors on target cells.
   d. often die before five years of age.
   e. both a and b.

10. Individuals with type II diabetes
   a. have an autoimmune disease in which the islets of Langerhans are attacked.
   b. are allergic to insulin.
   c. have an abnormally low number of insulin receptors on target cells.
   d. often die before five years of age.
   e. both a and b.

Matching

1. __________ Oxytocin
2. __________ ACTH
3. __________ Thyroxine
4. __________ Parathyroid hormone
5. __________ Somatotropin
6. __________ Melanocyte-stimulating hormone
7. __________ Vasopressin
8. __________ Second messengers

a. Stimulation of muscle and bone growth
b. Amplification of an original hormone signal
c. Water retention by the kidneys
d. The initiation of milk release in mothers
e. Regulation of blood calcium levels
f. Contains iodine
g. Stimulation of the adrenal gland to produce steroid hormones
h. Stimulation of color changes in reptiles

Completion

1. There are three classes of chemical messengers in the body: __________, __________, and __________.
2. Synthetic compounds that resemble testosterone are called the __________.
3. The adrenal gland is made of two parts: the __________ and the __________.
4. Insulin is produced by cells that are called __________.

Cool Places on the 'Net

Information about endocrinology and many disorders associated with the endocrine system can be obtained from the MedicalInfo Online at:

http://www.medicalinfo.com/Endocrinology.html

Information about diabetes can be found at the American Diabetes Association. Their address is:

http://www.diabetesnet.com/ada.html

Label the Art

Label the major glands of the endocrine system.

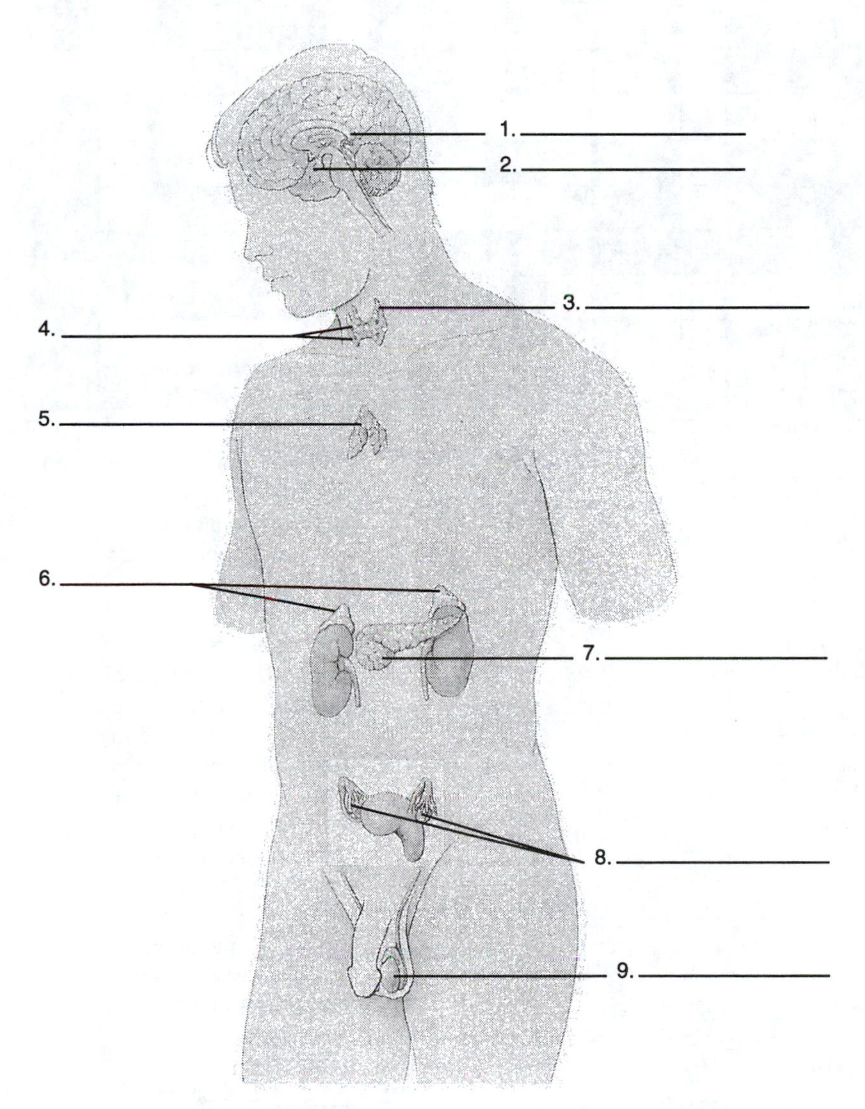

Crossword

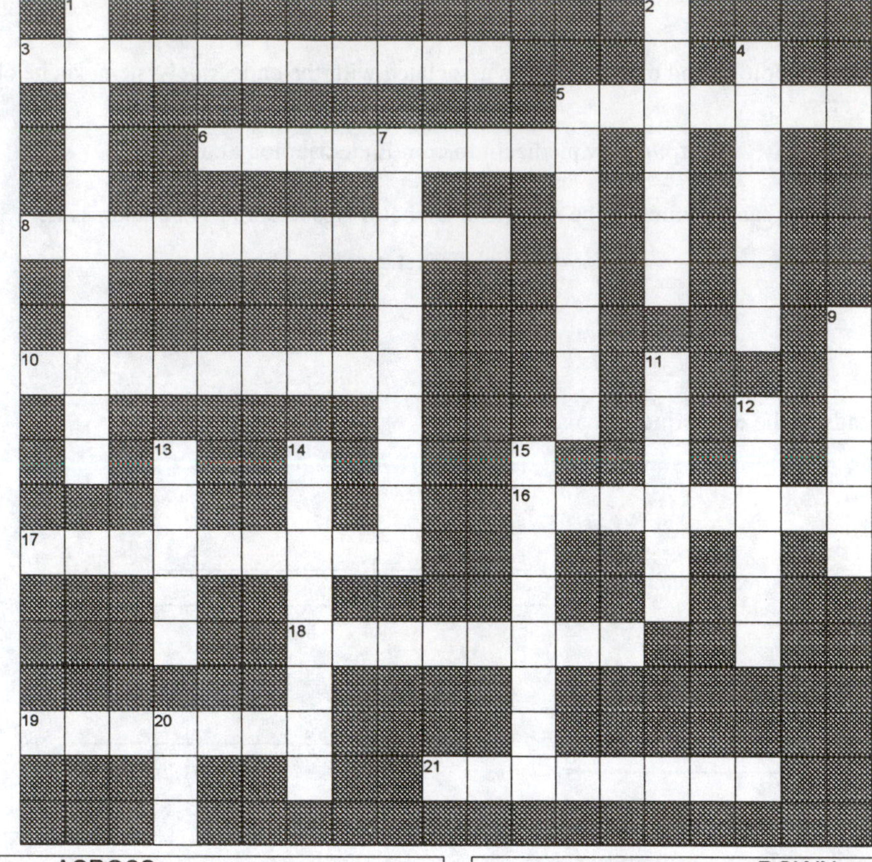

ACROSS

3. It is also called antidiuretic hormone
5. These glands are located above the kidneys
6. It stimulates the breasts to produce milk
8. The source of all steroid hormones
10. It promotes the breakdown of muscle proteins in metabolism
16. It initiates milk release in mothers
17. This gland hangs by a short stalk
18. Sweat glands are an example of these
19. These hormones act at the cell surface
21. These hormones are lipid-soluble

DOWN

1. This hormone regulates calcium levels in the blood
2. Long-lasting chemicals that are released from glands
4. Hormones are produced in one part of the body and act in ______ part of the body
5. Some athletes use these steroid hormones
7. A growth disorder
9. These messengers amplify the original signal
9. ________ messengers only work within cells
11. It produces aldosterone
12. This AMP is made from ATP
13. ______transmitters
14. There are two forms of this disease
15. _______-stimulating hormone is abreviated FSH
20. Short for thyrotropin-releasing hormone

25 Human Reproduction and Development

Chapter Concepts

Male gametes (sperm) are produced in the testes in very large numbers. At least 20 million must be delivered during intercourse for fertilization to be likely. Fertilization occurs in the fallopian tubes within a few days after egg release. When the developing embryo reaches the uterus, it embeds within the endometrium.

The hypothalamus regulates the reproductive cycle with a battery of hormones. A reproductive cycle takes about 30 days and ends with expulsion of the endometrium. The fertilized egg cleaves into a ball of cells called a blastomere. Cell movements during gastrulation form a hollow embryo with three primary tissues. By three weeks after fertilization, the body has a neural cord and somites.

The major organs of the body form in the fourth week. The limbs of the body assume their adult shape in the second month. Development is essentially complete eight weeks after fertilization.

The most effective nonsurgical methods of birth control involve using hormones or other chemicals to prevent egg maturation or embryo implantation.

Multiple Choice

1. The scrotum is located between the legs of the male rather than located internally because
   a. that is a safer location for the testes.
   b. the temperature is cooler there, allowing sperm to complete their development.
   c. the external location makes it easier to transport sperm to the penis during ejaculation.
   d. the higher temperatures that occur there allow the sperm to develop normally.
   e. both c and d.

2. Men with sperm counts of fewer than __________ sperm per milliliter are considered sterile.
   a. 600 million
   b. 250 million
   c. 100 million
   d. 50 million
   e. 20 million

3. Progesterone is produced
   a. only in postmenopausal women.
   b. by the ovaries.
   c. by the fallopian tubes.
   d. by the corpus luteum.
   e. just after menstruation.

4. A blastomere is
   a. the first opening that occurs in a blastula.
   b. a fluid-filled cavity.
   c. a cell in a morula.
   d. found in the mesoderm of the gastrula.
   e. an invagination on the surface of the embryo.

5. A blastocoel is
   a. the first opening that occurs in a blastula.
   b. a fluid-filled cavity.
   c. a cell in a morula.
   d. found in the mesoderm of the gastrula.
   e. an invagination on the surface of the embryo.

6. The miniature limbs of the embryo assume their adult shapes
   a. at three weeks of development.
   b. during the second month of development.
   c. during the third month of development.
   d. during the second trimester.
   e. during the third trimester.

7. Which of the following is the most effective form of birth control?
   a. condom
   b. birth-control pill
   c. rhythm method
   d. vasectomy
   e. coitus interruptus

Matching—Part 1

1. __________ Epididymis
2. __________ Seminiferous tubules
3. __________ Testosterone
4. __________ Sperm
5. __________ Vas deferens
6. __________ Scrotum
7. __________ Urethra
8. __________ Testes

a. Primarily used for storage of sperm
b. This has an acrosome
c. A sac that is about 3° C cooler than the rest of the body
d. Many tightly coiled tubes
e. Where the reproductive and urinary tracts meet
f. The sperm-producing organ
g. The maturation of sperm occurs here
h. A steroid hormone

Matching—Part 2

1. __________ Fallopian tube
2. __________ Follicle
3. __________ Cervix
4. __________ Ova
5. __________ Uterus
6. __________ Endometrium

a. An opening that is surrounded by a muscular ring
b. A mature egg
c. The outer layer is shed during menstruation
d. Cleavage of the zygote to the morula stage occurs here
e. Implantation of the zygote occurs here
f. The oocyte and a surrounding mass of tissue

Completion

1. Fertilization of an egg by a sperm results in the formation of a(n) __________ .
2. __________ is the hormone that causes ovulation in women.
3. During development, a membrane called the __________ interacts with uterine tissue to form the placenta, while another membrane called the __________ encloses the embryo.

Cool Places on the 'Net

The Medical Matrix on Embryology offers well-maintained links to guides, sites, journals, and organizations with information relating to embryology.

For information on sexual sciences, reproduction, and human relations go to:

http://www.zelacom.com/-hawthorn/humansxl.htm

Label the Art

A. Label the parts of the male reproductive system.

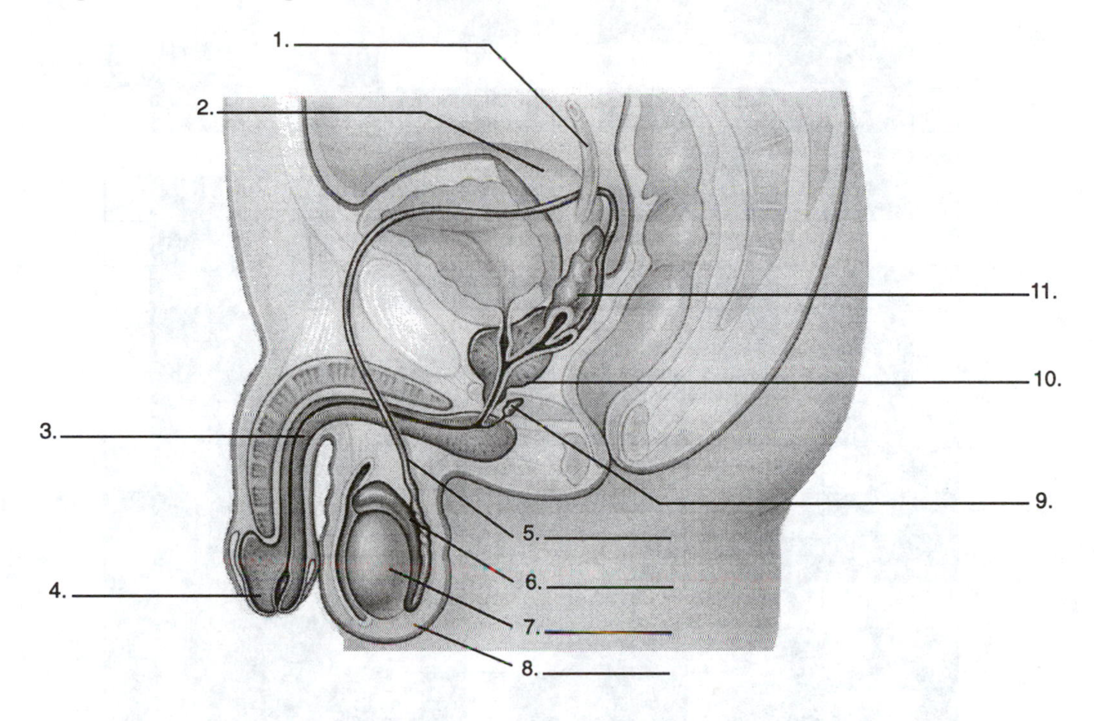

B. Label the parts of the female reproductive system.

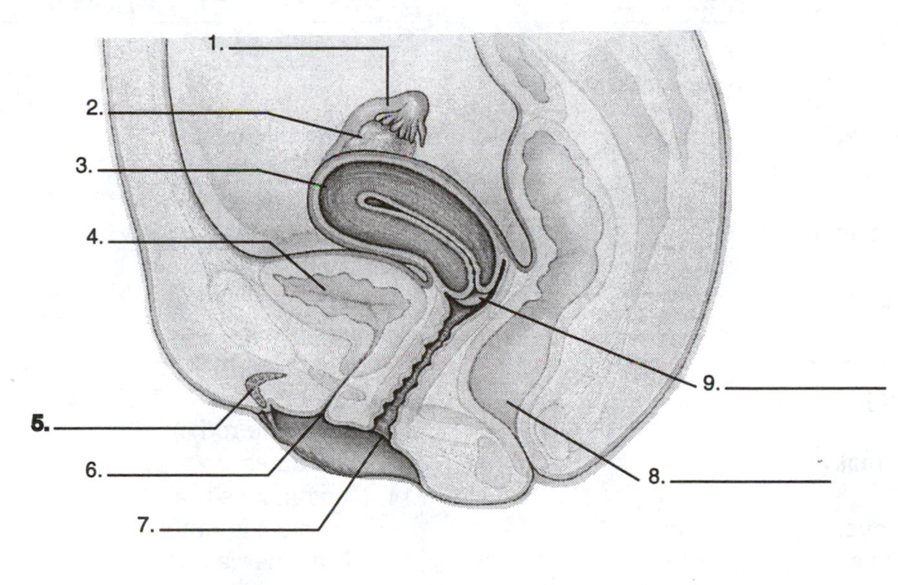

Crossword

ACROSS		DOWN	
1.	A female gamete	1.	Ejection of semen
2.	A long coiled tube where sperm mature	2.	The inner wall of the uterus
5.	A thin rubber bag	3.	A female cycle
6.	These tubes lead to the uterus	4.	_____ deferens
10.	The space within a tube	7.	Contains an enzyme that helps sperm digest a passageway for fertilization
11.	An opening in the uterus	8.	It initiates milk release
12.	A male sex hormone	9.	A flexible rod in chordates
13.	A cycle of egg maturation	14.	It nourishes the developing embryo
16.	It stimulates milk production	15.	The inner layer of cells
18.	The middle layer	17.	The fetus develops here
19.	Blocks of tissue		
20.	A developing human		

26 Ecosystems

Chapter Concepts

Ecology is the study of how organisms fit into and interact with their environment. Within a particular environment, many ecosystems may exist. An ecosystem is a self-sustaining group of organisms and the minerals, water, and weather that make up their habitat.

The flow of energy drives the growth and interaction of organisms within ecosystems. Producers are able to capture energy from sunlight by conducting photosynthesis, while consumers must eat plants or animals to obtain their energy. Ecologists categorize organisms into trophic or feeding levels based on how they obtain their energy. The complex flow of energy among trophic levels is called a food web.

Water cycles either by evaporation and condensation or by absorption and transpiration. The breaking up of molecules, the burning of wood and fossil fuels, and erosion are ways in which carbon cycles. Although nitrogen is abundant in organisms and the atmosphere, much of life relies on the ability of some bacteria to fix nitrogen.

The climate of a particular area is affected by the intensity of sun rays, air currents, elevation, and ocean currents. A rain shadow effect is caused by mountains, which force winds upwards, causing them to cool and release their moisture on the windward side of mountains. Ocean ecosystems consist of highly diverse coastal areas, open waters that contain plankton, and deep waters where little light penetrates. A biome is a climatically defined assemblage of organisms that occurs over a wide area.

Multiple Choice

1. Ecology is the study of
   a. how organisms interact with their environment.
   b. where we live.
   c. how organisms interact with each other.
   d. the different environments in the world.
   e. all of the above.

2. A variety of different bacteria and yeasts live on human skin. Together they are considered to be a(n)
   a. ecosystem.
   b. population.
   c. community.
   d. trophic level.
   e. biome.

3. Photosynthetic bacteria are in trophic level
   a. 1.
   b. 2.
   c. 3.
   d. all of the above.
   e. none of the above.

4. Molds that break down leaves and other dead organic material on the forest floor are in trophic level
   a. 1.
   b. 2.
   c. 3.
   d. all of the above.
   e. none of the above.

5. A lion that feeds on zebras is in trophic level
   a. 1.
   b. 2.
   c. 3.
   d. all of the above.
   e. none of the above.

6. The primary consumers of ecosystems are
   a. herbivores.
   b. carnivores.
   c. omnivores.
   d. the insects.
   e. both a and d.

7. Primary productivity refers to the
   a. rate of photosynthesis in plants and photosynthetic bacteria as compared to the rate of plant consumption by herbivores.
   b. total amount of light energy converted to organic compounds in a given area per unit time.
   c. rate of decomposition by the detritivores.
   d. total biomass of the photosynthetic organisms in the ecosystem.
   e. none of the above.

8. In the earth's ecosystems, energy
   a. is recycled and is never really lost.
   b. flows from producers to consumers and back to producers.
   c. flows in one direction, from producers to consumers.
   d. is produced at all trophic levels.
   e. both a and d.

9. A consequence of cutting down forests is
   a. that water is no longer returned to the atmosphere over the area of the forest.
   b. the loss of animal habitats.
   c. the production of high-quality agricultural land.
   d. all of the above.
   e. both a and b.

10. Burning wood
    a. releases oxygen into the atmosphere.
    b. consumes carbon dioxide.
    c. destroys the carbon that was in the wood.
    d. releases carbon into the atmosphere.
    e. releases nitrogen into the atmosphere.

11. Nitrogen fixation is accomplished by
    a. bacteria in the atmosphere.
    b. bacteria in the soil and in plant root nodules.
    c. plants.
    d. the fungi.
    e. all of the above.

12. Why did their corn grow better when Native Americans added a fish to the soil at planting time?
    a. The fish added carbon to the soil.
    b. The fish added "fixed" nitrogen to the soil.
    c. The fish added phosphorus and sulfur to the soil.
    d. The fish provided the vitamins that the young plant required.
    e. All of the above.

13. Pollution of lakes by commercial detergents that contain phosphates
    a. kills the bacteria there and causes a breakdown in the food chain.
    b. encourages the growth of bacteria and an increase in the fish population.
    c. encourages the growth of algae, which leads to the suffocation of fish and other animals.
    d. raises the pH.
    e. lowers the pH.

14. Ocean currents are determined by
    a. proximity to land.
    b. underwater geography.
    c. atmospheric circulation.
    d. the season.
    e. none of the above.

15. Deep sea waters below 300 meters
    a. harbor a rich variety of life including the red algae.
    b. are rich in photoplankton.
    c. consist mostly of plant life.
    d. harbor limited kinds of life, some of which are very strange.
    e. both b and c.

16. Oligotrophic lakes are
    a. only found at high latitudes.
    b. saltwater lakes.
    c. rich in organic matter and nutrients.
    d. scarce in organic matter and nutrients.
    e. rich in plant life.

Matching

1. __________ Tundra
2. __________ Tropical rain forest
3. __________ Grasslands
4. __________ Chaparral
5. __________ Savannas
6. __________ Deciduous forests
7. __________ Desert

a. The richest ecosystems on earth are found here
b. Many of the animal residents live in burrows
c. Dry climates that border tropics; grasslands
d. Low-growing, spiny plants; Mediterranean climate
e. Deer, rabbits, raccoons, and squirrels
f. Permafrost may be a meter deep
g. They are also called prairies

Completion

1. That portion of the ecosystem that includes soil, water, and weather is called the __________ .
2. A community and a habitat together are called a(n) __________ .
3. The path from producer through carnivore is called a(n) __________ .
4. Plants lose water by __________ and water reenters the atmosphere by __________ .
5. Excessive nutrients in an aquatic ecosystem lead to rapid, uncontrolled growth and that is referred to as __________ .
6. Intermediate areas that lie between freshwater and saltwater are called __________ .
7. A(n) __________ is a terrestrial ecosystem that occurs over a broad area.

Cool Places on the 'Net

The Society for Ecological Restoration is an international professional membership organization that promotes ecological restoration as a means of sustaining biodiversity on earth. Their address is:

http://nabalu.flas.ufl.edu/ser/SERhome.html

Key Word Search

B	E	N	M	C	A	R	N	I	V	O	R	E	S	K
W	I	O	C	E	N	F	L	J	P	Y	U	V	R	F
U	R	I	H	M	O	O	T	B	E	T	A	A	E	R
S	I	T	A	O	I	O	R	I	R	I	G	P	M	S
T	A	A	P	I	T	D	O	O	M	N	I	O	U	U
R	R	R	A	B	A	C	P	M	A	U	A	R	S	O
E	P	I	R	A	C	H	H	A	F	M	T	A	N	U
S	J	P	R	D	I	A	I	S	R	M	E	T	O	D
E	K	S	A	C	H	I	C	S	O	O	F	I	C	I
D	F	N	L	J	P	N	L	N	S	C	A	O	V	C
I	F	A	L	L	O	V	E	R	T	U	R	N	E	E
M	K	R	S	E	R	O	V	I	T	I	R	T	E	D
E	F	T	M	O	T	S	E	R	O	F	N	I	A	R
S	X	X	L	E	U	F	L	I	S	S	O	F	H	U
K	Z	I	N	T	E	R	T	I	D	A	L	R	H	P

Biomass
Biome
Carnivores
Chaparral
Community
Consumers
Deciduous
Deserts
Detritivores
Eutrophic lake
Eutrophication
Evaporation
Fall overturn
Food chain
Fossil fuel
Intertidal
Permafrost
Prairie
Rain forest
Semidesert
Taiga
Transpiration
Trophic level

Crossword

ACROSS	DOWN

ACROSS

1. There are a lot of these in the southwestern part of the United States
3. They eat plants
6. A small body of water
7. Evergreen spiny shrubs and low plants
8. It would be a really big planet without these
10. Using it over again
12. A pretty big body of water
13. This kind of shadow caused Death Valley
14. The collection of organisms that live in a particular area
18. Ocean ______ can be modified by land masses
19. To wear away
20. Fussy eaters they are not
21. This word was first used by Haekel

DOWN

2. This is how plants lose water
4. This is how water reenters the atmosphere
5. Dry, tropical grasslands
6. Plants are this
9. They eat plant eaters
10. Humans do plenty of this
11. Saltwater and freshwater
15. Much energy is ______
16. Water and minerals are part of the ______ habitat
17. A food _____ is very complex
18. This cycle is begun by plants photosynthesizing

27 Living in Ecosystems

Chapter Concepts

A population is a group of individuals of the same species that live together. The innate rate at which a population grows is a function of the organism's physiology. The actual growth rate of a population is affected by birth and death, immigration and emigration, the carrying capacity, and population density.

Different species within ecosystems have coevolved to compete with, interact with, and benefit from each other. In symbiotic relationships, two species interact such that only one benefits or both benefit from the interaction.

Within an ecosystem, organisms can potentially occupy a range of habitats and behave in many different ways. Organisms usually only occupy part of their fundamental niche due to competition. In predator-prey interactions, organisms may advertise their toxicity using aposematic coloration or hide their palatability using cryptic coloration. Often palatable species mimic the coloration of toxic ones to fool predators.

The number of different species, or species richness, in an ecosystem determines its biological diversity. Ecosystems with biodiversity are more stable, while at the same time, they may also have points of vulnerability. A larger size and equatorial latitude promote biodiversity in an ecosystem.

The scale at which humans cause disturbance is often too fast and too large for ecosystems to accommodate. Minimizing pollution, the clear-cutting of forests, and the introduction of exotic species will help to preserve our ecosystems.

Multiple Choice

1. Population dispersion may be
   a. uniform.
   b. clumped.
   c. random.
   d. all of the above.
   e. both a and b.

2. When referring to the capacity of a population to grow, *r* equals the
   a. number of new individuals in the population—the number of individuals leaving the population.
   b. carrying capacity of the population.
   c. population growth rate.
   d. dispersal of the population.
   e. reproductive potential of the population.

3. The rapid growth of bacteria that occurs after inoculation into agar in a petri plate is called a(n)
   a. bloom.
   b. lawn.
   c. exponential growth.
   d. accelerated growth.
   e. activated growth.

4. Carrying capacity is the number of
   a. offspring that an individual can produce.
   b. offspring that an individual can raise to maturity.
   c. individuals that can be supported in a place indefinitely.
   d. organisms that a carnivore consumes in its lifetime.
   e. none of the above.

5. A rise in the incidence of tuberculosis in crowded prisons is an example of a(n)
   a. survivorship effect.
   b. density-dependent effect.
   c. dispersion effect.
   d. limiting factor.
   e. density-independent effect.

6. The age distribution of a population influences
   a. the carrying capacity.
   b. survivorship curve.
   c. intrinsic rate of increase.
   d. dispersion.
   e. emigration of individuals.

7. In order to protect themselves from herbivores plants may
   a. contain toxic substances.
   b. contain unpleasant tasting substances.
   c. have thorns.
   d. have spines.
   e. all of the above.

8. Lichens are an example of
   a. parasitism.
   b. commensalism.
   c. mutualism.
   d. synergism.
   e. none of the above.

9. Bees that pollinate flowers are an example of
   a. parasitism.
   b. commensalism.
   c. mutualism.
   d. synergism.
   e. none of the above.

10. Aphids living on roses are an example of
    a. parasitism.
    b. commensalism.
    c. mutualism.
    d. synergism.
    e. none of the above.

11. Brightly colored poison dart frogs are an example of
    a. cryptic coloration.
    b. aposematic coloration.
    c. Batesian mimicry.
    d. Müllerian mimicry.
    e. none of the above.

12. The monarch butterfly is an example of
    a. cryptic coloration.
    b. aposematic coloration.
    c. Batesian mimicry.
    d. Müllerian mimicry.
    e. none of the above.

13. Tomato worms are exactly the same color as the plants that they feed on. This is an example of
    a. cryptic coloration.
    b. aposematic coloration.
    c. Batesian mimicry.
    d. Müllerian mimicry.
    e. none of the above.

14. Lichens growing on the surface of rocks provide an example of
    a. Batesian mimicry.
    b. primary succession.
    c. secondary succession.
    d. a climax community.
    e. a keystone species.

15. The growth that occurs after a forest fire has occurred is an example of
    a. regrowth.
    b. primary succession.
    c. secondary succession.
    d. a climax community.
    e. keystone speciation.

Completion

1. That part of earth where life can be supported is called the __________ .
2. The rate at which a population increases when there are no limits on its rate of growth is called the __________ or the __________ .
3. The environment in which an organism lives and how it competes with other organisms is called a(n) __________ .

Cool Places on the 'Net

For information on conservation ecology, contact the Ecological Society of America at:

http://journal.biology.carleton.ca/

For a list of World Wide Web sites related to ecology, start at:

http://biomserv.univ-lyonl.fr/Ecology-WWW.html

Key Word Search

S	U	C	C	E	S	S	I	O	N	I	C	H	E	P
E	L	B	A	T	E	F	I	L	Y	F	T	V	E	O
I	D	P	I	H	S	N	O	I	T	A	L	E	R	P
C	L	E	A	R	C	U	T	T	I	N	G	T	J	U
E	T	M	S	M	S	I	T	I	S	A	R	A	P	L
P	I	H	S	R	O	V	I	V	R	U	S	P	E	A
S	Z	M	S	I	L	A	S	N	E	M	M	O	C	T
C	I	T	O	I	B	M	Y	S	V	F	R	S	O	I
I	C	O	M	P	E	T	I	T	I	O	N	E	S	O
T	C	I	T	P	Y	R	C	P	D	Z	H	M	Y	N
O	C	O	L	O	R	A	T	I	O	N	E	A	S	F
X	C	D	I	S	P	E	R	S	I	O	N	T	T	F
E	R	E	H	P	S	O	I	B	B	C	Z	I	E	S
M	S	I	L	A	U	T	U	M	T	S	L	C	M	Z
D	E	N	S	I	T	Y	C	D	J	A	Z	I	W	P

Aposematic
Biodiversity
Biosphere
Clear cutting
Coloration
Commensalism
Competition
Cryptic

Density
Dispersion
Ecosystem
Exotic species
Life table
Mutualism
Niche
Parasitism

Population
Relationship
Size
Succession
Survivorship
Symbiotic

Crossword

ACROSS	DOWN

ACROSS

2. Population ______ is the number of individuals in a unit area
5. A warning strategy
6. A good deal for both parties
8. Population _____ is the number of individuals in a population
9. Richness of life
10. These species can cause a problem in a new ecosystem
13. A struggle between two organisms
18. This rate tells the number of individuals that are dead at any age
19. Many butterflies and moths do this type of mimicry
20. A plant defense mechanism

DOWN

1. Two species evolving together
3. Species replacement
4. One species benefits, the other neither benefits nor is it harmed
7. How an organism lives in an ecosystem
11. Individuals of the same species living together
12. The part of earth that can support life
14. Individuals leaving a population
15. A _____ species is critical in an ecosystem
16. This capacity is the number of individuals that can be supported in an area
17. ______ potential

28 Planet Under Stress

Chapter Concepts

Air and water pollution result mainly from increased industrialization, oil spills, the use of agricultural chemicals, and unwise disposal of wastes. Acid rain, precipitation polluted by sulfuric acid originating from factories, kills trees and lakes by lowering the pH level.

Chlorofluorocarbons (CFCs) used in coolants, aerosols, and foaming agents are eating the earth's ozone, which exposes life on earth to harmful ultraviolet radiation. An increase in carbon dioxide in the atmosphere is responsible for increased global warming, or the greenhouse effect.

Antipollution laws, pollution taxes, and economic evaluations are underway to reduce pollution. Nuclear power provides an alternative source of energy to the burning of coal and oil, although the challenges of operation must be overcome. We must slow down our use of topsoil and groundwater, resources that took over thousands of years to accumulate. With the destruction of tropical and temperate rain forests, organisms with potentially vital roles in the ecosystem are being lost. The alarmingly high human population growth rate is at the core of many environmental problems.

Environmental problems have been overcome through assessment, analysis, education, and following through. You can do your part by recycling, educating others about the environment, voting, and writing letters.

Multiple Choice

1. Which of the following can cause acid rain?
   a. soil erosion
   b. industrial smokestacks
   c. a decrease in the ozone layer
   d. clear-cutting forests
   e. none of the above

2. Widespread effects on the worldwide ecosystem are called
   a. disasters.
   b. global change.
   c. beneficial.
   d. succession.
   e. both c and d.

3. Why does it take so long for plastics to break down in nature?
   a. Ozone in the atmosphere slows down the process.
   b. Bacteria and fungi are unable to break down plastics.
   c. Plastics become stronger with age.
   d. Plastics can only be broken down by incineration.
   e. None of the above.

4. __________ released when coal is burned contributes to the production of acid rain.
   a. Nitrogen
   b. Carbon
   c. Sulfur
   d. Phosphorus
   e. Carbon dioxide

5. Lakes with a pH of __________ cannot support most life.
   a. 10
   b. 8
   c. 7
   d. 5
   e. none of the above

6. A serious drawback to the use of nuclear power is
   a. the disposal of radioactive materials.
   b. that it cannot provide very much energy.
   c. the shortage of radioactive materials.
   d. the difficulty in guarding nuclear power plants
   e. both a and d.

7. To solve environmental problems we must
   a. assess the situation.
   b. educate the public.
   c. predict the consequences of environmental intervention.
   d. vote.
   e. all of the above.

Completion

1. An oil tanker called the __________ was involved in a huge oil spill in Alaska in 1989.
2. A class of chemicals called __________ are destroying the ozone layer.
3. Every __________% drop in atmospheric ozone leads to a __________% increase in the incidence of skin cancers.
4. The __________ effect prevents heat from radiating into space.
5. __________ helps prevent the waste of resources.
6. Plowing for the planting of crops has resulted in a loss of __________ .
7. Water trapped beneath the soil in porous rock is called __________ .
8. Humans first came to North America about __________ years ago.
9. By the year 2000, about __________% of the world's population will be living in tropical or subtropical countries.

Cool Places on the 'Net

For information on how to prevent pollution, see the National Pollution Prevention Center for Higher Education page. The address is:

http://www2.snre.umich.edu/nppc/

For information about air pollution, go to:

http://www-wilson.ucsd.edu/education/airpollution/airpollution.html

Appendix

Answers to Chapter Questions

Chapter 1

Multiple Choice

1. c
2. b
3. e
4. b
5. c
6. b
7. e
8. a
9. e
10. b
11. c
12. c
13. a
14. b

Completion

1. hypothesis
2. metabolism
3. DNA, genes
4. tissues
5. a community, the physical environment
6. Charles Darwin, *On The Origin of Species*

Key Word Search

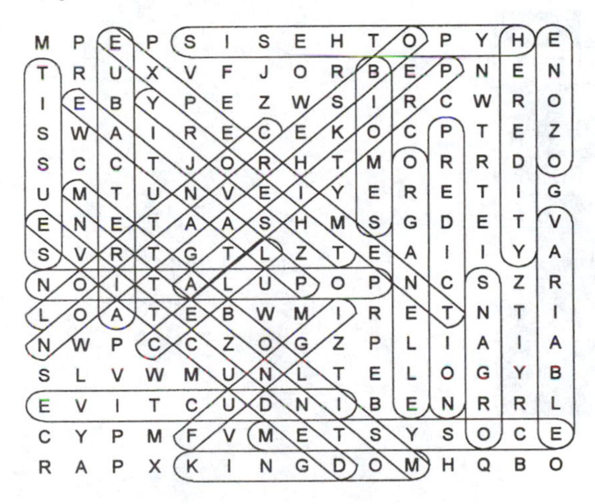

Crossword

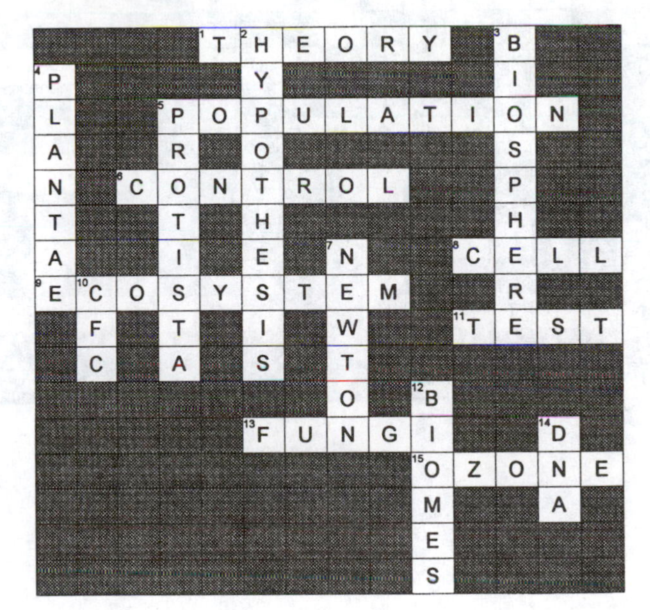

Chapter 2

Multiple Choice

1. b
2. d
3. c
4. d
5. a
6. b
7. b and c
8. a
9. d
10. c
11. a
12. a
13. b
14. d
15. c
16. d
17. a
18. d
19. e
20. a

Completion

1. extraterrestrial origins, special creation, evolution
2. cohesion, adhesion
3. acidic, basic
4. polysaccharides
5. thymine, guanine

Label the Art

1. hydroxyl group
2. carbonyl group
3. carboxyl group
4. amino group
5. sulfhydryl group
6. phosphate group

Crossword

[1]W				[2]S			[3]C			[4]R			[5]C					
E		[6]O		T			[7]A	D	H	E	S	I	O	N				[8]B
I		X		E			R			D			H			[9]D	N	A
G		I		R			B			U			E					S
[10]H	Y	D	R	O	P	H	O	B	I	C		[11]I	S	O	T	[12]O	P	E
T		A		I			N			T			I			R		
		T		D		[13]E		[14]A	C	I	D		[15]O	X	Y	G	E	N
		I				L				O			N			A		
	[16]C	O	V	A	L	E	N	T		N						N		[17]H
		N				C			[18]S			[19]P	E	P	T	I	D	E
						T			T							C		L
[20]I			[21]H	Y	D	R	O	G	E	N		[22]I	O	N				I
O						O			A			C						X
[23]N	E	U	T	[24]R	O	N			[25]M	O	L	E	C	U	L	E		
I				N														
C				A														

Chapter 3

Multiple Choice

1. d
2. e
3. a
4. a
5. d
6. b
7. d
8. e
9. d
10. a

Matching

1. b
2. c
3. i
4. f
5. h
6. d
7. a
8. g
9. j
10. e

Completion

1. 5, 20
2. volume, surface area
3. cystic fibrosis
4. cytoskeleton
5. actin, tubulin

Label the Art

A. Animal cell
   1. Golgi complex
   2. microvilli
   3. plasma membrane
   4. centrioles
   5. smooth endoplasmic reticulum
   6. mitochondrion
   7. rough endoplasmic reticulum
   8. nucleus
   9. cytoskeleton
   10. nucleolus
   11. nuclear envelope
   12. lysosome

B. Chloroplast
   1. stroma
   2. thylakoid
   3. lamella
   4. outer membrane
   5. inner membrane
   6. granum

Key Word Search

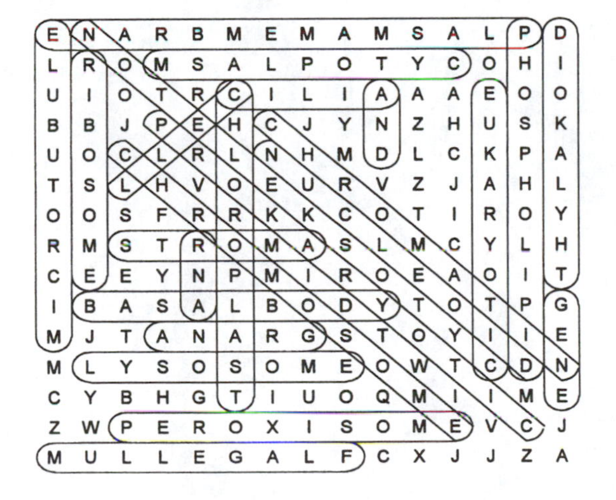

Crossword

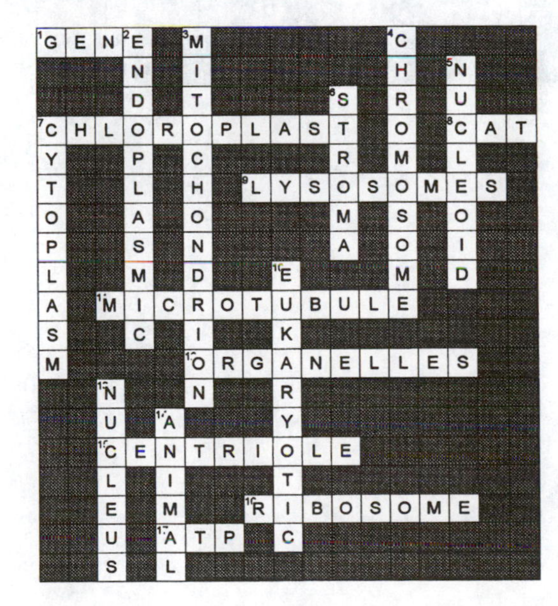

Chapter 4

Multiple Choice

1. c
2. a
3. a
4. a
5. b
6. b
7. e
8. b
9. c
10. c
11. a
12. c
13. d
14. c
15. c
16. b
17. c
18. a
19. e
20. d

Completion

1. active transport, ATP
2. sodium-potassium pump
3. binary fission

Label the Art

1. metaphase
2. interphase
3. telophase
4. daughter cells
5. anaphase
6. prophase

Crossword

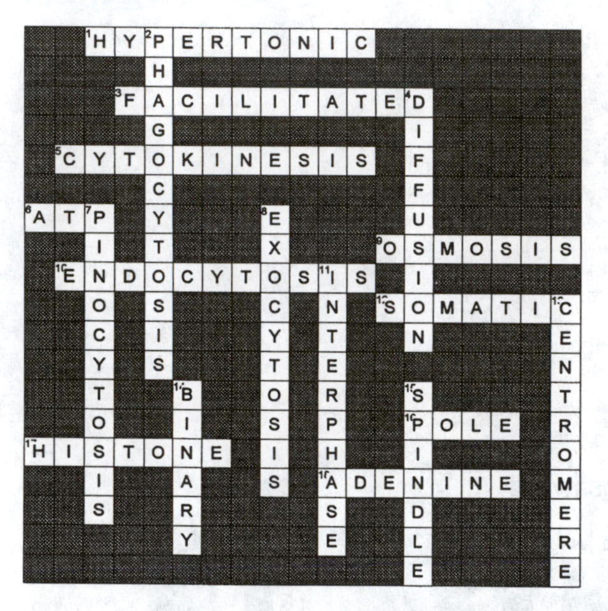

Chapter 5

Multiple Choice

1. e
2. a
3. c
4. e
5. b
6. a
7. d
8. c
9. b
10. c
11. b
12. d
13. a
14. c
15. d

Completion

1. endergonic, exergonic
2. binding site
3. coenzyme
4. 700, 680
5. in the stomata of the chloroplast, carbohydrates

Label the Art

1. crista
2. matrix
3. inner membrane

Crossword

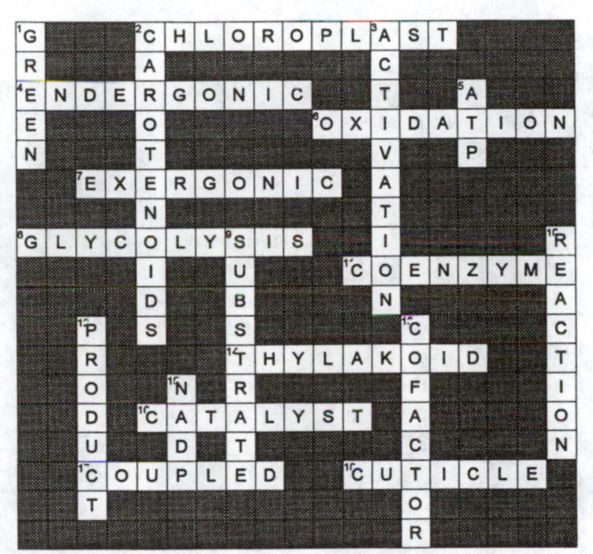

Chapter 6

Multiple Choice

1. e
2. c
3. b
4. a
5. c
6. c
7. b
8. d
9. e
10. c
11. c
12. b
13. c
14. d
15. b
16. d
17. b
18. e
19. a

Completion

1. genes, alleles
2. *Aa, Ba, Ab, Bb*
3. probability
4. test cross
5. pleiotrophic
6. haploid, diploid
7. karyotype

Label the Art

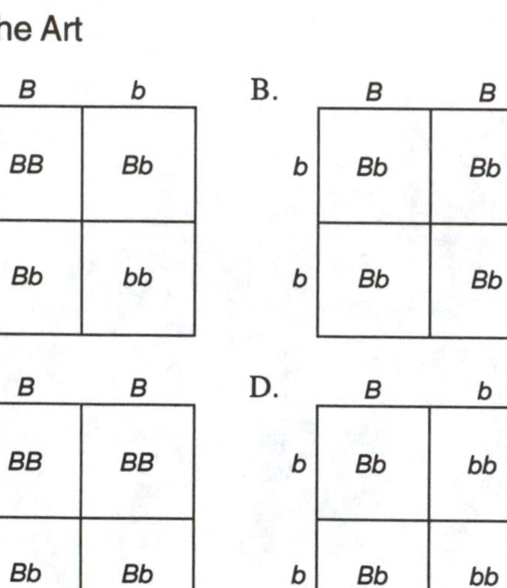

Crossword

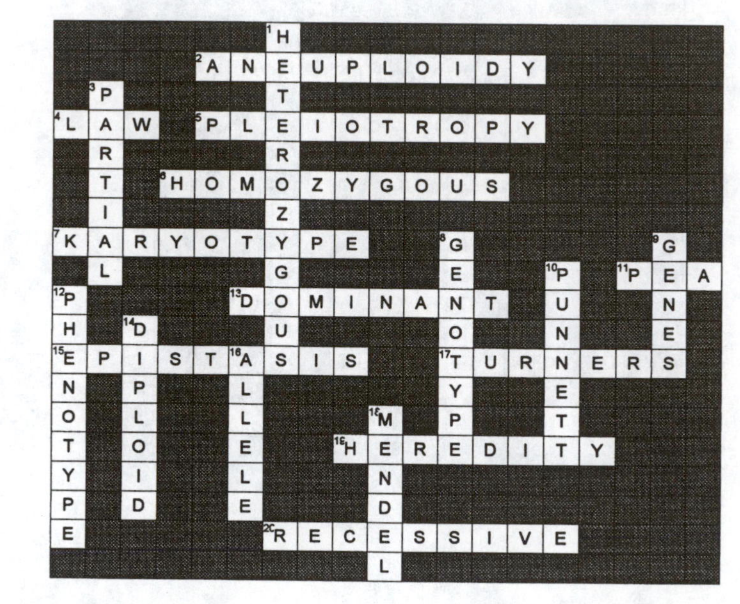

Chapter 7

Multiple Choice

1. a
2. c
3. b
4. c
5. d
6. c
7. b
8. c
9. c
10. d
11. a
12. b
13. e
14. b
15. d
16. c
17. d

Completion

1. replication fork
2. semiconservative
3. operon
4. exons, introns
5. mutagens

Label the Art

1. mRNA transcribed from DNA
2. tRNA binds amino acids
3. loaded tRNA bonds to ribosomes
4. growing peptide chain
5. completed protein

Crossword

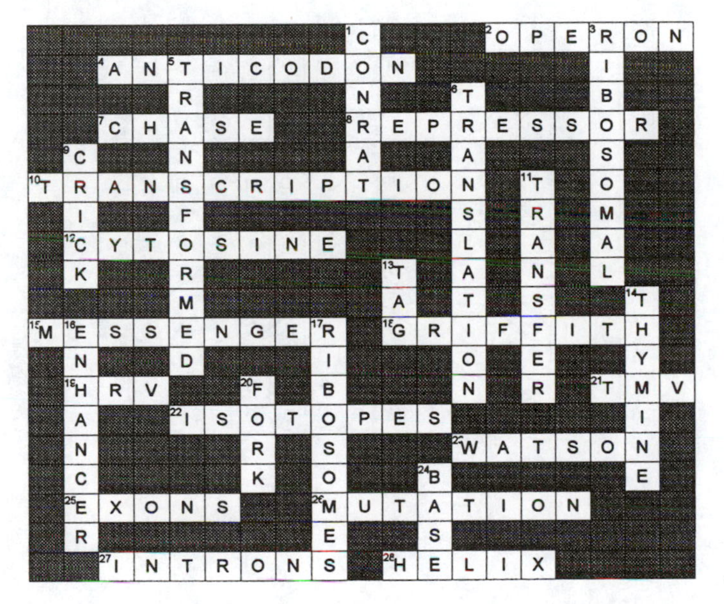

Chapter 8

Multiple Choice

1. c
2. a
3. b
4. c
5. c
6. d
7. b
8. e
9. d

Completion

1. genetic engineering
2. sticky ends
3. cleaving DNA, recombining DNA, cloning, screening for the new DNA in the recipient organism
4. probe
5. nitrogen fixation, increases

Matching

1. d
2. e
3. f
4. h
5. a
6. b
7. g
8. c

Key Word Search

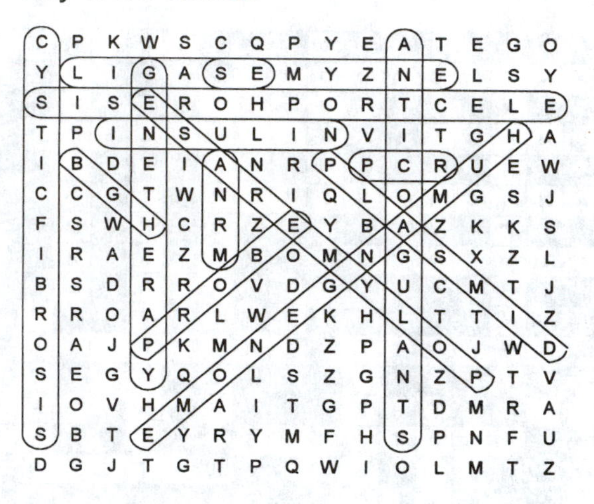

Chapter 9

Multiple Choice

1. e
2. d
3. c
4. b
5. a
6. c
7. a
8. c
9. e
10. a
11. e

Matching—Part 2

1. e
2. c
3. g
4. i
5. d
6. f
7. b
8. h
9. a

Matching—Part 1

1. c
2. d
3. a
4. e
5. b

Completion

1. *Archaeopteryx,* 150 million years ago
2. radio-isotopic dating, less than 50,000 years old
3. inbreeding
4. stabilizing
5. species

Key Word Search

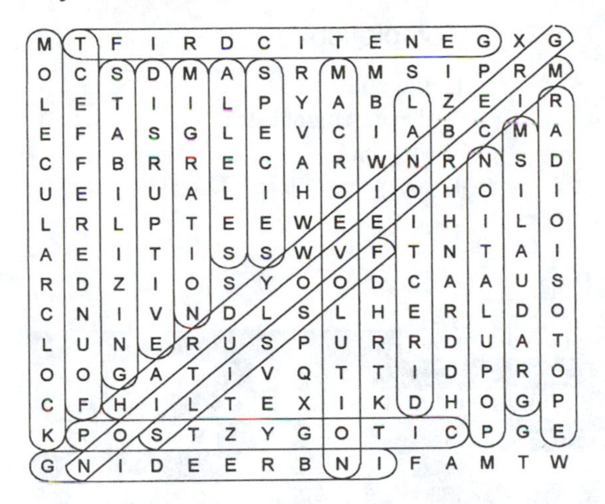

Crossword

Chapter 10

Multiple Choice

1. a
2. b
3. d
4. d
5. e
6. b
7. d
8. b
9. a
10. a
11. e
12. a
13. d
14. c
15. b

Completion

1. family
2. asexually
3. outgroup
4. 1.5 million

Key Word Search

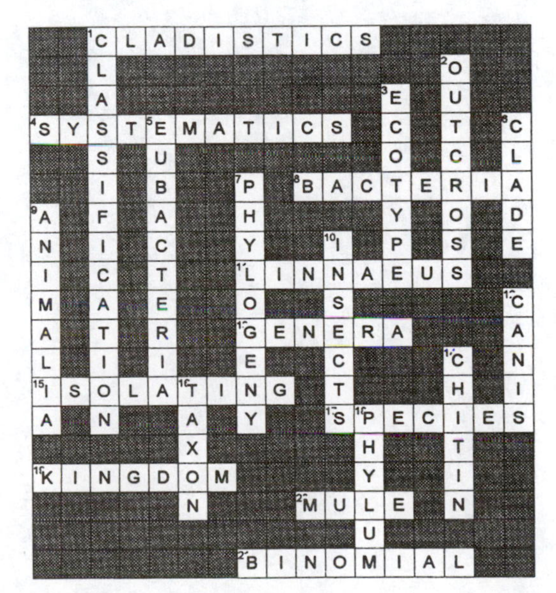

Crossword

Chapter 11

Multiple Choice

1. b
2. c
3. d
4. e
5. d
6. b
7. a
8 e
9. c

Matching

1. g
2. f
3. h
4. j
5. i
6. c
7. d
8. a
9. e
10. b

Completion

1. flagella
2. thermoacidophiles
3. viruses
4. receptor protein CD4

Label the Art

A. Bacterial cell
   1. capsule
   2. ribosomes
   3. cytoplasm
   4. plasma membrane
   5. pili
   6. cell wall
   7. DNA
   8. flagellum

B. Bacteriophage
   1. head
   2. neck
   3. tail
   4. base plate
   5. tail fiber

Crossword

		[1]C	A	P	S	U	L	E									[2]I	
		Y								[3]C		[4]M		[5]G	R	A	M	
		A		[6]E	N	D	O	S	P	O	R	E	S				M	
		N								N		T		[7]V	I	R	U	S
	[8]C	O	C	C	U	S				J		H					N	
		B								U		[9]A	R	C	H	A	E	
[10]T		A		[11]G		[12]H		[13]S		G		N						
[14]B	A	C	T	E	R	I	O	P	H	A	G	E						
		T		N		V		I		T			[15]F	[16]I	X			
		E		E				K		I				N				
		R			[17]H	[18]E	T	E	R	O	C	Y	S	T				
		I				N		S		N				R				
	[19]W	A	L	L		V			[20]B		[21]B			O				
						E			I		A			N				
[22]F	L	A	G	E	L	L	A		L		[22]C	A	P	S	I	D		
L						O			L		I							
U					[24]S	P	I	R	I	L	L	A						
						E			O		L							
									N		I							

Chapter 12

Multiple Choice

1. b
2. d
3. b
4. d
5. a
6. c
7. c
8. d
9. e
10. d

Matching

1. i
2. e
3. g
4. j
5. a
6. h
7. d
8. b
9. f
10. c

Completion

1. alternation of generations
2. forams
3. *Anopheles,* sporozoite

Label the Art

A. *Euglena*
   1. flagellum
   2. stigma
   3. contractile vacuole
   4. pellicle
   5. nucleus
   6. chloroplast

B. The life cycle of *Plasmodium*
   1. mosquito injects sporozoites
   2. sporozoites
   3. stages in liver
   4. merozoites
   5. stages in red blood cells
   6. certain merozoites develop into gametocytes
   7. gametocytes
   8. gametocytes are ingested by mosquito
   9. sporozoites form within mosquito

Crossword

Chapter 13

Multiple Choice

1. a
2. d
3. b
4. c
5. b
6. e
7. c
8. a
9. c
10. d
11. d
12. a
13. c
14. d
15. b
16. c
17. b
18. e
19. a

Completion

1. colonial
2. development
3. hyphae, mycelium
4. zygomycetes
5. dikaryon
6. mushroom

Label the Art

1. spores
2. sporangium
3. rhizoid
4. hypha
5. gametangia
6. zygospore
7. germinating zygospore
8. sporangiophore
9. sporangium

Crossword

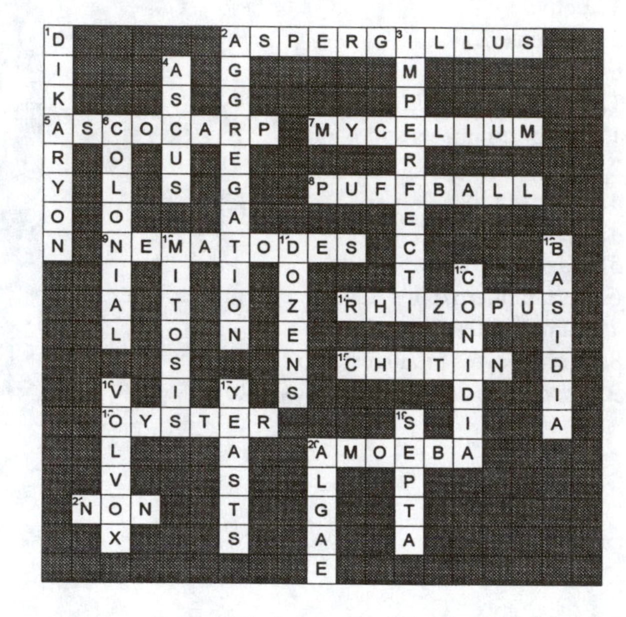

Chapter 14

Multiple Choice

1. c
2. c
3. e
4. c
5. d
6. b
7. a
8. d
9. a
10. c
11. b
12. c

Matching

1. c
2. e
3. a
4. f
5. g
6. b
7. d

Completion

1. autotrophs
2. pollen grain
3. Coniferophyta

Label the Art

A. Generalized plant life cycle
   1. gametophyte
   2. *n*
   3. antheridia
   4. archegonia
   5. sperm
   6. egg
   7. zygote
   8. embryo
   9. sporophyte
   10. sporangia
   11. spore mother cell
   12. tetrads of spores
   13. spore

B. Parts of the flower
   1. sepal
   2. petal
   3. stamen
   4. stigma
   5. style
   6. ovary

Crossword

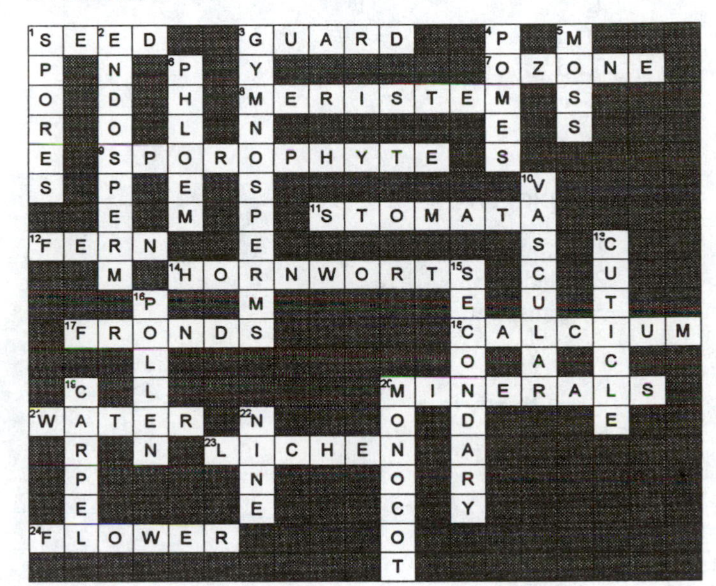

Chapter 15

Multiple Choice

1. a
2. b
3. e
4. b
5. b
6. c
7. d
8. b
9. d
10. b
11. a

Matching—Part 2

1. e
2. a
3. d
4. c
5. b

Completion

1. shoot, root
2. blade
3. transpiration

Label the Art

1. shoot
2. root
3. apical meristem
4. terminal bud
5. leaf
6. axillary bud
7. primary growth zone
8. secondary growth zone

Matching—Part 1

1. f
2. g
3. j
4. h
5. b
6. d
7. a
8. e
9. i
10. c

Crossword

T	R	A	C	H	E	I	D				C				F			
			O					M	E	S	O	P	H	Y	L	L		
M			H	A	I	R	S				M				O			
E			E								P				R			
R			S			P	I	T	H		A	B	S	C	I	S	I	C
I		S	I	E	V	E					N				G			A
S			O			R			E		I		B	E	E	S		M
T	R	A	N	S	P	I	R	A	T	I	O	N			N			B
E						C			H		N							I
M		N				Y			Y					C				U
S		E		S		C			L					O				M
		C		P		L		P	E	R	I	D	E	R	M			
P	E	T	I	O	L	E			N					O				
		A		N			S	T	E	M		T		L				
		R		G			H					U		L				
				Y			O			D	E	R	M	A	L			
							O		C			G						
				R	O	O	T		A	U	X	I	N					
									P			D						

Chapter 16

Multiple Choice

1. a
2. b
3. c
4. c
5. d
6. e
7. d
8. b
9. c
10. d
11. e
12. e
13. b
14. d
15. c
16. a

Completion

1. Arthropoda, Mollusca, Chordata
2. Ctenophora
3. Acoelomate, Platyhelminthes
4. circulation, movement, organ function
5. primary induction
6. Arthropoda
7. chitin, a site for internal muscle attachment
8. chelicerates, mandibulates
9. head, thorax, abdomen
10. two, one

Label the Art

1. cheliped
2. eye
3. cephalothorax
4. abdomen
5. telson
6. uropod
7. swimmerets
8. walking legs
9. antenna
10. antennule

Crossword

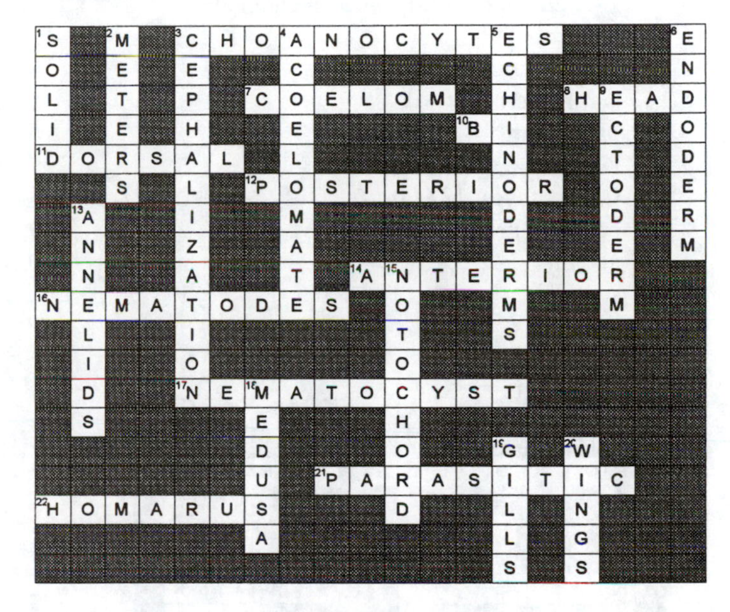

Chapter 17

Multiple Choice

1. d
2. e
3. a
4. b
5. b
6. e
7. b
8. c
9. e
10. c
11. b
12. a
13. d
14. d
15. b

Matching

1. d
2. b
3. a
4. e
5. c

Completion

1. 470 million, 100 million
2. 248, 265, terrestrial plants and animals
3. Paleozoic, 96
4. Triassic, Jurassic, Cretaceous
5. warm and moist
6. placoderms
7. lateral line system
8. Osteichthyes, 18,000
9. countercurrent flow
10. alveoli
11. septum

Label the Art

1. yolk sac
2. chorion
3. albumin
4. egg shell
5. air space
6. egg membrane
7. allantois
8. chorioallantoic membrane
9. amnion
10. embryo

Crossword

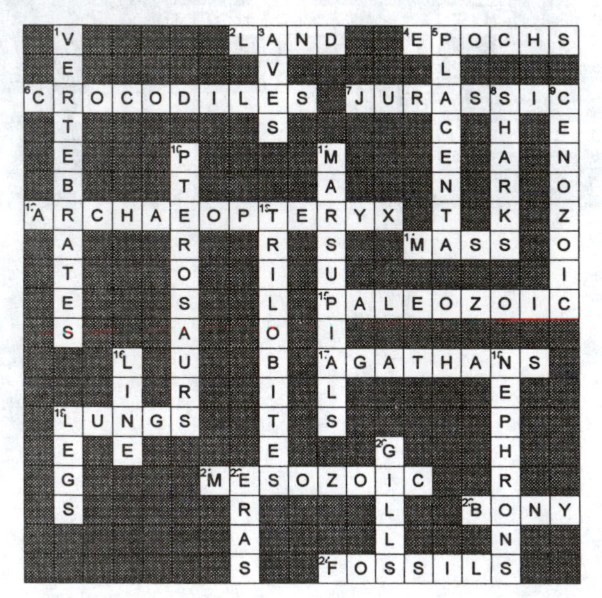

Chapter 18

Multiple Choice

1. d
2. d
3. a
4. c
5. e
6. e
7. e
8. b
9. b

Matching

1. d
2. c
3. e
4. b
5. a

Completion

1. *The Descent of Man*
2. grasping hands
3. lemurs
4. 3%
5. Java
6. tools, fire

Key Word Search

S	U	T	C	E	R	E	O	M	O	H	Y	R	Z	K
K	S	N	E	I	P	A	S	O	M	O	H	M	H	K
C	H	C	Y	P	Y	L	U	L	V	M	M	L	O	N
L	O	M	A	R	U	M	C	G	L	O	A	J	U	G
S	M	P	B	N	S	V	U	Z	V	H	E	D	E	J
V	I	R	V	C	I	S	C	P	T	A	I	M	F	B
F	N	O	B	E	W	N	R	R	B	B	Y	C	D	E
B	O	S	Y	Z	Q	E	E	O	V	I	I	S	P	I
Y	I	I	C	K	M	D	Y	S	S	L	O	Y	U	T
Y	D	M	U	O	N	A	M	G	N	I	K	E	P	S
M	S	I	L	A	D	E	P	I	B	S	C	K	O	C
W	I	A	E	L	E	K	R	E	V	H	J	N	M	H
V	R	N	A	M	A	V	A	J	S	A	J	O	I	L
S	D	S	E	T	A	M	I	R	P	W	V	M	R	C
F	C	Z	H	B	D	R	W	T	R	X	I	Z	R	V

Crossword

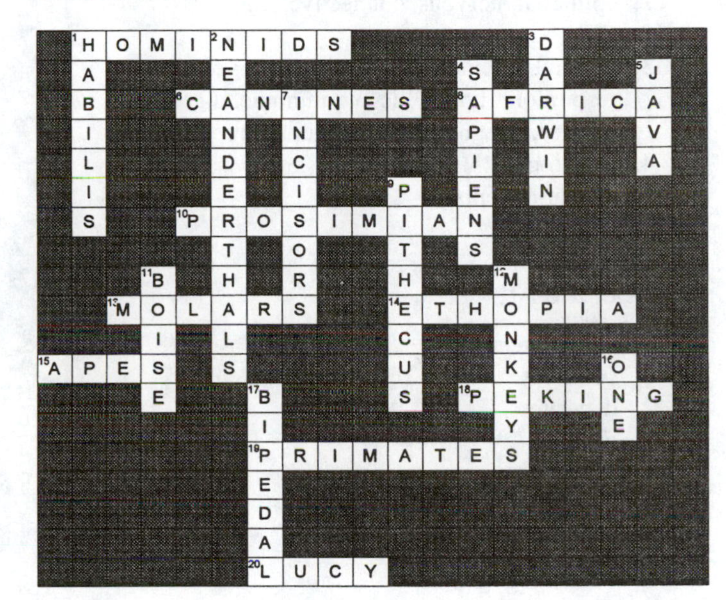

Chapter 19

Multiple Choice

1. a
2. d
3. d
4. b
5. c
6. e
7. b
8. d
9. b
10. c

Matching

1. b
2. c
3. d
4. e
5. a

Completion

1. epithelial, nervous, connective, muscle
2. 100
3. skeletal, circulatory, endocrine, nervous, respiratory, lymphatic and immune, digestive, urinary, integumentary, muscular, reproductive
4. erythrocytes
5. microfilaments
6. 206, 80, 126

Crossword

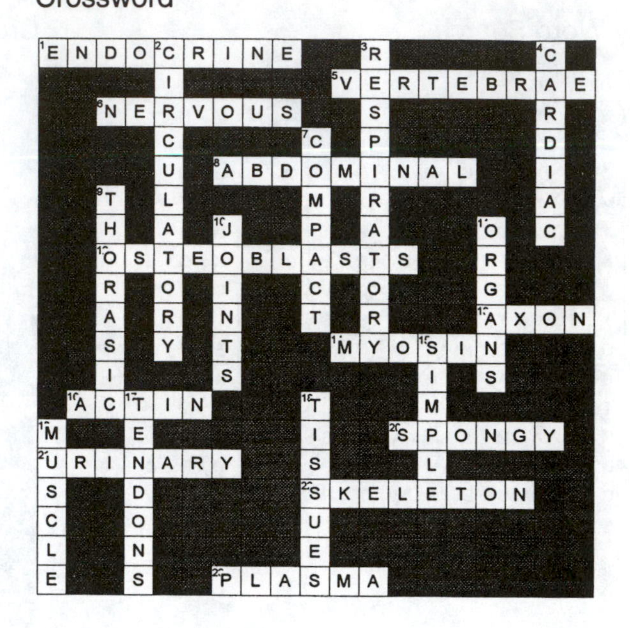

Label the Art

Parts of the neuron

1. cell body
2. dendrites
3. nucleus
4. axon

Chapter 20

Multiple Choice

1. c
2. d
3. a
4. a
5. d
6. b
7. a
8. e
9. c
10. d
11. b
12. e
13. a
14. d
15. b

Matching

1. a
2. e
3. b
4. d
5. c

Completion

1. heart, blood vessels, blood
2. metabolites and wastes, salts and ions, proteins
3. atherosclerosis
4. tumors, carcinomas

Label the Art

A. Cardiovascular system
   1. superior vena cava
   2. axillary artery
   3. inferior vena cava
   4. aorta
   5. femoral artery and vein
   6. popliteal artery
   7. pulmonary artery
   8. heart

B. Respiratory system
   1. lungs
   2. trachea
   3. pharynx
   4. nasal cavity
   5. oral cavity
   6. larynx
   7. bronchus

Crossword

Chapter 21

Multiple Choice

1. c
2. e
3. a
4. c
5. e
6. c
7. d
8. b
9. c
10. b

Matching

1. d
2. e
3. b
4. a
5. c

Completion

1. 9.3, 4.1
2. stomach and small intestine
3. mix food with saliva
4. esophagus
5. epithelial lining
6. insulin

Label the Art

A. Human digestive system
   1. salivary glands
   2. liver
   3. gallbladder
   4. large intestine
   5. pharynx
   6. esophagus
   7. stomach
   8. pancreas
   9. small intestine
   10. rectum

B. Human tooth
   1. enamel
   2. dentin
   3. pulp
   4. gingiva
   5. bone
   6. root canal

Crossword

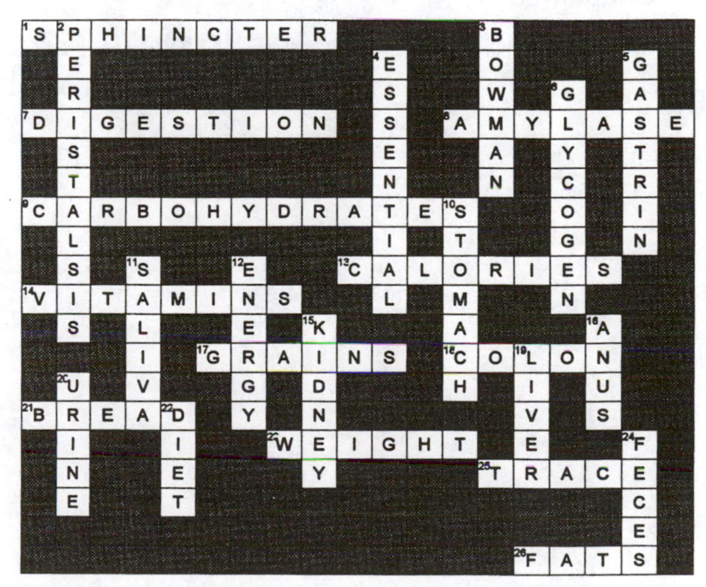

Chapter 22

Multiple Choice

1. d
2. b
3. c
4. c
5. e
6. d
7. d
8. b
9. d
10. c
11. a
12. d
13. d

Matching

1. b
2. f
3. d
4. a
5. c
6. e

Completion

1. skin
2. epidermis, dermis, subcutaneous layer
3. inflammatory response
4. macrophages, helper T cells
5. antibodies
6. piggyback vaccine
7. autoimmune

Label the Art

1. epidermis
2. dermis
3. subcutaneous layer
4. hair shaft
5. sweat gland pore
6. capillary
7. sweat gland duct
8. sebaceous gland
9. arector pili muscle
10. hair follicle
11. nerve fiber
12. sweat gland
13. fat cells
14. blood vessels

Crossword

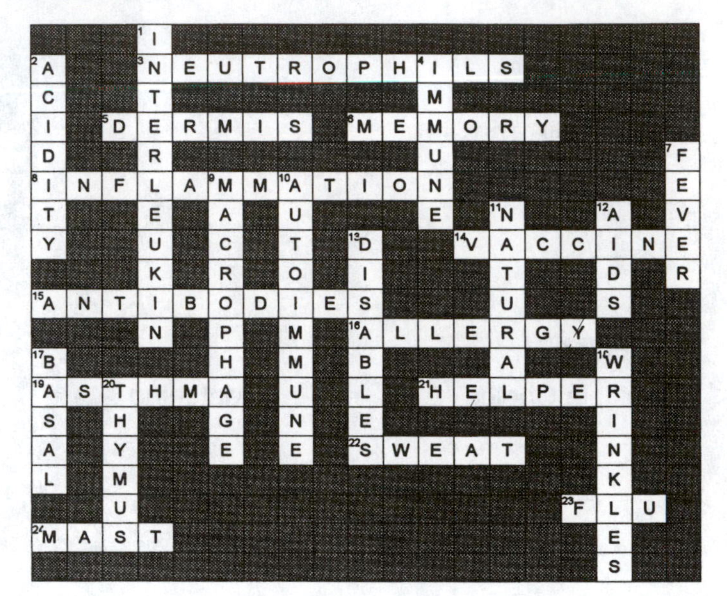

Chapter 23

Multiple Choice

1. c
2. c
3. b
4. c
5. a
6. b
7. a
8. b

Matching—Part 1

1. d
2. e
3. b
4. a
5. c

Matching—Part 2

1. c
2. d
3. e
4. b
5. a

Completion

1. hypothalamus, cerebral cortex
2. sympathetic nervous system
3. presynaptic membrane, postsynaptic membrane
4. chemically gated potassium channel
5. sonar
6. cornea

Label the Art

1. corpus striatum
2. thalamus
3. cerebral cortex
4. corpus callosum
5. skull
6. pineal gland
7. cerebellum
8. pons
9. spinal cord
10. medulla oblongata
11. reticular formation
12. pituitary gland
13. hypothalamus

Crossword

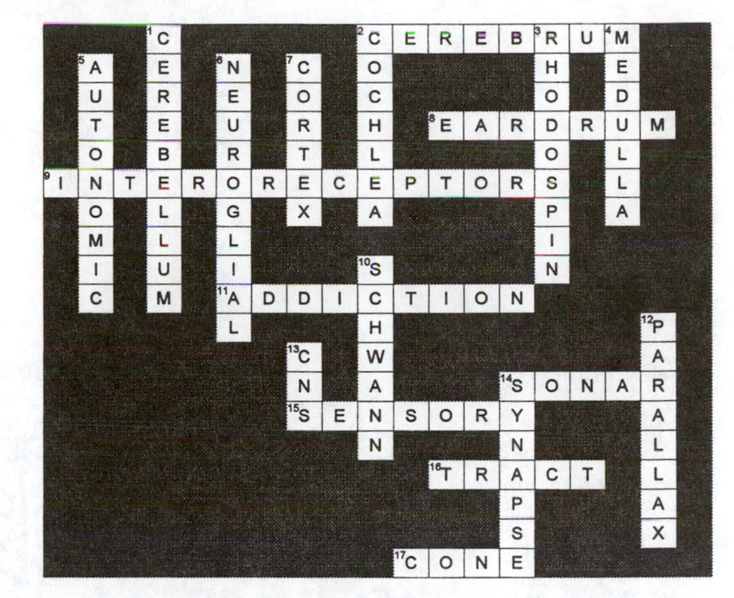

Chapter 24

Multiple Choice

1. e
2. c
3. d
4. c
5. b
6. a
7. d
8. a
9. a
10. c

Matching

1. d
2. g
3. f
4. e
5. a
6. h
7. c
8. b

Completion

1. second messengers, neurotransmitters, hormones
2. anabolic steroids
3. medulla, cortex
4. islets of Langerhans

Label the Art

1. pineal gland
2. pituitary gland
3. thyroid gland
4. parathyroid glands
5. thymus
6. adrenal glands
7. pancreas
8. ovaries
9. testes

Crossword

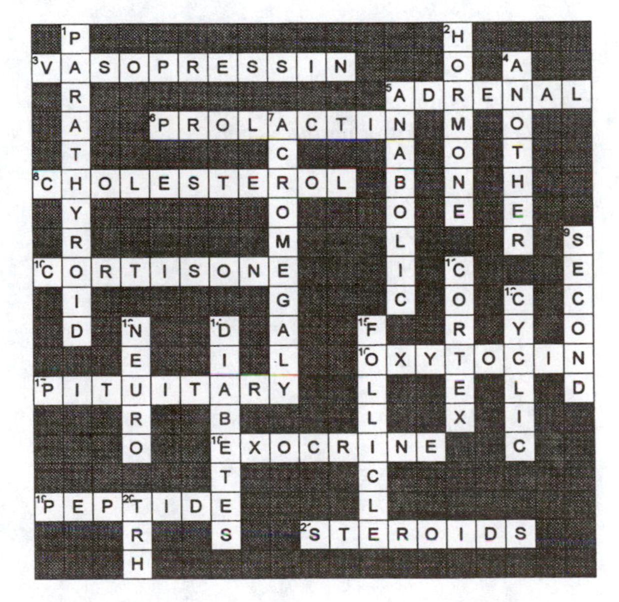

Chapter 25

Multiple Choice

1. b
2. e
3. d
4. c
5. b
6. b
7. d

Matching—Part 1

1. g
2. d
3. h
4. b
5. a
6. c
7. e
8. f

Matching—Part 2

1. d
2. f
3. a
4. b
5. e
6. c

Completion

1. zygote
2. luteinizing hormone
3. chorion, amnion

Label the Art

A. Male reproductive system
   1. ureter
   2. bladder
   3. urethra
   4. penis
   5. vas deferens
   6. epididymis
   7. testis
   8. scrotum
   9. Cowper's gland
   10. prostate gland
   11. seminal vesicle

B. Female reproductive system
   1. fallopian tube
   2. ovary
   3. uterus
   4. bladder
   5. clitoris
   6. urethra
   7. vagina
   8. rectum
   9. cervix

Crossword

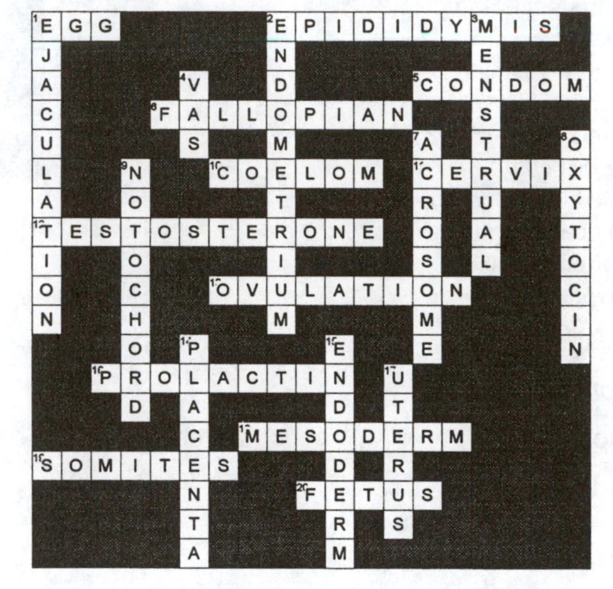

Chapter 26

Multiple Choice

1. e
2. c
3. a
4. d
5. c
6. a
7. b
8. c
9. e
10. d
11. b
12. b
13. c
14. c
15. d
16. d

Matching

1. f
2. a
3. g
4. d
5. c
6. e
7. b

Completion

1. physical habitat
2. ecosystem
3. food chain
4. transpiration, evaporation
5. eutrophication
6. estuaries
7. biome

Key Word Search

B	E	N	M	C	A	R	N	I	V	O	R	E	S	K
W	I	O	C	E	N	F	L	J	P	Y	U	V	R	F
U	R	I	H	M	O	O	T	B	E	T	A	A	E	R
S	I	T	A	O	I	O	R	I	R	I	G	P	M	S
T	A	A	P	I	T	D	O	O	M	N	I	O	U	U
R	R	R	A	B	A	C	P	M	A	U	A	R	S	O
E	P	I	R	A	C	H	H	A	F	M	T	A	N	U
S	J	P	R	D	I	A	I	S	R	M	E	T	O	D
E	K	S	A	C	H	I	C	S	O	O	F	I	C	I
D	F	N	L	J	P	N	L	N	S	C	A	O	V	C
I	F	A	L	L	O	V	E	R	T	U	R	N	E	E
M	K	R	S	E	R	O	V	I	T	I	R	T	E	D
E	F	T	M	O	T	S	E	R	O	F	N	I	A	R
S	X	X	L	E	U	F	L	I	S	S	O	F	H	U
K	Z	I	N	T	E	R	T	I	D	A	L	R	H	P

Crossword

[1]D	E	S	E	R	[2]T	S		[3]H	[4]E	R	B	I	V	O	R	E	[5]S	
					R				V								A	
[6]P	O	N	D		A		[7]C	H	A	P	A	R	R	A	L		V	
R					N				P								A	
O					S				O								N	
[8]D	E	[9]C	O	M	P	O	S	E	R	S		[10]C	Y	C	L	I	N	G
U		A			I				A			O					A	
C		R			R				T			N			[11]E		S	
E		N		[12]L	A	K	E		I			S			S			
[13]R	A	I	N		T			[14]C	O	M	M	U	N	I	T	Y		
S		V			I		[15]L		N			M			U			
		O			O		O			[16]P		E			A			[17]W
[18]C	U	R	R	E	N	T	S			H		R		[19]E	R	O	D	E
A		E					T			Y		S			I			B
R		S								S					E			
B							[20]O	M	N	I	V	O	R	E	S			
O										C								
N										A								
							[21]E	C	O	L	O	G	Y					

Chapter 27

Multiple Choice

1. d
2. a
3. c
4. c
5. b
6. c
7. e
8. c
9. c
10. a
11. b
12. b
13. a
14. b
15. c

Completion

1. biosphere
2. innate capacity for increase or biotic potential
3. niche

Key Word Search

S	U	C	C	E	S	S	I	O	N	I	C	H	E	P
E	L	B	A	T	E	F	I	L	Y	F	T	V	E	O
I	D	P	I	H	S	N	O	I	T	A	L	E	R	P
C	L	E	A	R	C	U	T	T	I	N	G	T	J	U
E	T	M	S	M	S	I	T	I	S	A	R	A	P	L
P	I	H	S	R	O	V	I	V	R	U	S	P	E	A
S	Z	M	S	I	L	A	S	N	E	M	M	O	C	T
C	I	T	O	I	B	M	Y	S	V	F	R	S	O	I
I	C	O	M	P	E	T	I	T	I	O	N	E	S	O
T	C	I	T	P	Y	R	C	P	D	Z	H	M	Y	N
O	C	O	L	O	R	A	T	I	O	N	E	A	S	F
X	C	D	I	S	P	E	R	S	I	O	N	T	T	F
E	R	E	H	P	S	O	I	B	B	C	Z	I	E	S
M	S	I	L	A	U	T	U	M	T	S	L	C	M	Z
D	E	N	S	I	T	Y	C	D	J	A	Z	I	W	P

Crossword

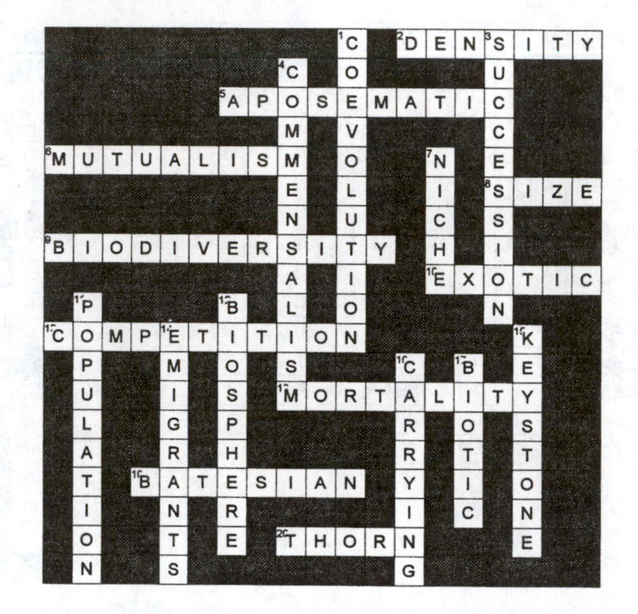

Chapter 28

Multiple Choice

1. b
2. b
3. b
4. c
5. d
6. e
7. e

Completion

1. *Exxon Valdez*
2. chlorofluorocarbons
3. 1, 6
4. greenhouse
5. recycling
6. topsoil
7. groundwater
8. 12,000 to 13,000
9. 60